Advances in Experimental Medicine and Biology

Neuroscience and Respiration

Volume 837

For further volumes:
http://www.springer.com/series/13457

Mieczyslaw Pokorski
Editor

Neurotransmitter Interactions and Cognitive Function

Editor
Mieczyslaw Pokorski
Institute of Psychology
University of Opole
Poland

ISSN 0065-2598 ISSN 2214-8019 (electronic)
ISBN 978-3-319-36370-7 ISBN 978-3-319-10006-7 (eBook)
DOI 10.1007/978-3-319-10006-7
Springer Cham Heidelberg New York Dordrecht London

Printed on acid-free paper

Springer is part of Springer Science+Business Media (www.springer.com)

Preface

This is a new book series entitled Neuroscience and Respiration, a subseries of Springer's renowned Advances in Experimental Medicine and Biology. The book volumes present contributions by expert researchers and clinicians in the field of pulmonary disorders. The chapters provide timely overviews of contentious issues or recent advances in the diagnosis, classification, and treatment of the entire range of pulmonary disorders, both acute and chronic. The texts are thought as a merger of basic and clinical research dealing with respiratory medicine, neural and chemical regulation of respiration, and the interactive relationship between respiration and other neurobiological systems such as cardiovascular function or the mind-to-body connection. In detail, topics include lung function, hypoxic lung pathologies, epidemiology of respiratory ailments, sleep-disordered breathing, imaging, and biomarkers. Other needful areas of interest are acute respiratory infections or chronic inflammatory conditions of the respiratory tract, exemplified by asthma and chronic obstructive pulmonary disease (COPD), or those underlain by still unknown factors, such as sarcoidosis, respiratory allergies, lung cancer, and autoimmune disorders involving the respiratory system.

The prominent experts will focus their presentations on the leading-edge therapeutic concepts, methodologies, and innovative treatments. Pharmacotherapy is always in the focus of respiratory research. The action and pharmacology of existing drugs and the development and evaluation of new agents are the heady area of research. Practical, data-driven options to manage patients will be considered. The chapters will present new research regarding older drugs, performed from a modern perspective or from a different pharmacotherapeutic angle. The introduction of new drugs and treatment approaches in both adults and children will be discussed. The problem of drug resistance, its spread, and deleterious consequences will be dealt with as well.

Lung ventilation is ultimately driven by the brain. However, neuropsychological aspects of respiratory disorders are still mostly a matter of conjecture. After decades of misunderstanding and neglect, emotions have been rediscovered as a powerful modifier or even the probable cause of various somatic disorders. Today, the link between stress and respiratory health is undeniable. Scientists accept a powerful psychological connection that can directly affect our quality of life and health span. Psychological approaches, by decreasing stress, can play a major role in the development and course of respiratory disease, and the mind-body techniques can aid in their treatment.

Neuromolecular aspects relating to gene polymorphism and epigenesis, involving both heritable changes in the nucleotide sequence and functionally relevant changes to the genome that do not involve a change in the nucleotide sequence, leading to respiratory disorders will also be tackled. Clinical advances stemming from basic molecular and biochemical research are but possible if the research findings are "translated" into diagnostic tools, therapeutic procedures, and education, effectively reaching physicians and patients. All that cannot be achieved without a multidisciplinary, collaborative, "bench-to-bedside" approach involving both researchers and clinicians, which is the essence of the book series Neuroscience and Respiration.

The societal and economic burden of respiratory ailments has been on the rise worldwide leading to disabilities and shortening of life span. COPD alone causes more than three million deaths globally each year. Concerted efforts are required to improve this situation, and part of those efforts are gaining insights into the underlying mechanisms of disease and staying abreast with the latest developments in diagnosis and treatment regimens. It is hoped that the books published in this series will fulfill such a role by assuming a leading role in the field of respiratory medicine and research and will become a source of reference and inspiration for future research ideas.

Titles appearing in Neuroscience and Respiration will be assembled in a novel way in that chapters will first be published online to enhance their speedy visibility. Once there are enough chapters to form a book, the chapters will be assembled into complete volumes. At the end, I would like to express my deep gratitude to Mr. Martijn Roelandse and Ms. Tanja Koppejan from Springer's Life Sciences Department for their genuine interest in making this scientific endeavor come through and in the expert management of the production of this novel book series.

Opole, Poland Mieczyslaw Pokorski

Volume 6: Neurotransmitter Interactions and Cognitive Function

A host of neurotransmitters and neuroactive substances underlies respiratory regulation in health and disease. The centerpiece of investigations regarding adaptation to hypoxia and sensorial perception has been the dopaminergic system. It is now clear that a complex interaction among various neuroactive substances, rather than a single one, forms the basis of respiratory changes. The research on neurotransmitter interactions provides the knowledge of how the brain functions and a new level of understanding of mind-to-body connection, which opens up avenues for novel therapeutic interventions.

Contents

Advs Exp. Medicine, Biology - Neuroscience and Respiration (2015) 6: 1–8
DOI 10.1007/5584_2014_86

Published online: 14 October 2014

Inherited Disorders of Brain Neurotransmitters: Pathogenesis and Diagnostic Approach

Krystyna Szymańska, Katarzyna Kuśmierska, and Urszula Demkow

Abstract

Neurotransmitters (NTs) play a central role in the efficient communication between neurons necessary for normal functioning of the nervous system. NTs can be divided into two groups: small molecule NTs and larger neuropeptide NTs. Inherited disorders of NTs result from a primary disturbance of NTs metabolism or transport. This group of disorders requires sophisticated diagnostic procedures. In this review we discuss disturbances in the metabolism of tetrahydrobiopterin, biogenic amines, γ-aminobutyric acid, foliate, pyridoxine-dependent enzymes, and also the glycine-dependent encephalopathy. We point to pathologic alterations of proteins involved in synaptic neurotransmission that may cause neurological and psychiatric symptoms. We postulate that synaptic receptors and transporter proteins for neurotransmitters should be investigated in unresolved cases. Patients with inherited neurotransmitters disorders present various clinical presentations such as mental retardation, refractory seizures, pyramidal and extrapyramidal syndromes, impaired locomotor patterns, and progressive encephalopathy. Every patient with suspected inherited neurotransmitter disorder should undergo a structured interview and a careful examination including neurological, biochemical, and imaging.

K. Szymańska
Department of Experimental and Clinical Neuropathology, Medical Research Center, Polish Academy of Sciences, Warsaw, Poland

Department of Child Psychiatry, Medical University of Warsaw, 24 Marszalkowska St., 00-576 Warsaw, Poland

K. Kuśmierska
Department of Laboratory Diagnostics and Clinical Immunology, Medical University of Warsaw, 24 Marszalkowska St., 00-576 Warsaw, Poland

Department of Newborn Screening Laboratory, Institute of Mother and Child, Warsaw, Poland

U. Demkow (✉)
Department of Laboratory Diagnostics and Clinical Immunology, Medical University of Warsaw, 24 Marszalkowska St., 00-576 Warsaw, Poland
e-mail: urszula.demkow@litewska.edu.pl

Keywords
Biogenic amines • γ-aminobutyric acid • Glycine • Folic acid • Neurotransmitters • Pediatric neurotransmitter diseases • Pyridoxine • Tetrahydrobiopterin

1 Introduction

Neurotransmitters (NTs) play a central role in the nervous system homeostasis, movement, and behavior. Efficient communication between large numbers of neurons is necessary for normal functioning of the nervous system. A crucial mechanism of neuronal interaction involves the release of NTs into a synaptic cleft and binding to specific receptors. An understanding of synthesis pathways of NTs, their receptors and transporter proteins, locations, and interactions is central to brain function in normal and pathological conditions. An array of NTs must be functionally integrated to facilitate the enormous dynamics within the nervous system. Functional status of the central nervous system (CNS) is based on a highly regulated balance between excitatory and inhibitory circuits.

NTs can be divided into two groups: small molecule NTs and larger neuropeptide NTs. Glutamate, γ-amino butyric acid (GABA), glycine, and biogenic amines (dopamine, norepinephrine, serotonin, and histamine) belong to the former category (Table 1) (Pearl et al. 2006). Biogenic amines are transported into synaptic vesicles by the vesicular monoamine transporter (VMAT); dopamine is transported from the synapse to cytosol by dopamine transporter (DAT), and serotonin by serotonin transporter (SERT) (Lin et al. 2011). Reuptake of GABA at GABAergic synapses is performed by GABA transporter 1 (GAT1) (Anderson et al. 2010). The process of neurotransmission consists of synthesis of a neurotransmitter by a presynaptic neuron, accumulation and release of it into the synaptic cleft, binding to, and activation of, the receptor, reuptake from the synapse, and degradation (Pearl et al. 2006). Two cofactors play an important role in NTs metabolism – tetrahydrobiopterin and pyridoxine. The active form of vitamin B_6 pyridoxine-5-phosphate (PLP) is required for the synthesis of dopamine, serotonin, and histamine. PLP plays also a role in the metabolism of GABA and is a cofactor of the P-protein in the glycine cleavage system (Clayton 2006; Pearl et al. 2006). Inherited disorders of NTs belong to the group of neurometabolic syndromes

Table 1 Neurotransmitters and their role in the central nervous system

Biogenic amines	Dopamine	Regulation of motor activity and in integrative aspects of behavior, cognitive functions, and attention-related processes
	Norepinephrine	Regulation of vigilance, reactivity, and executive function
	Serotonin (5-hydroxytryptamine)	Modulation of tonic of nervous system activity, feeding behavior, and circadian rhythms. Precursor of the melatonin
	Histamine	Control of biological rhythms, behavioral state, energy metabolism, thermoregulation, fluid balance, and reproduction
Amino acids	Glutamate	Excitatory in the central nervous system. Regulation of sensorimotor functions, cognitive functions, emotions, and memory
	γ-amino butyric acid (GABA)	Inhibitory in the central nervous system
	Glycine	Inhibitory in the brainstem and spinal cord, co-agonist of glutamate on NMDA (excitatory) receptors
	Acetylcholine	Regulation of all motor activity, alertness, cognitive functions, emotions, and memory

related to the primary disturbance of NT metabolism or transport. This group of disorders requires sophisticated diagnostic procedures.

2 Neurotransmitters and Their Metabolites

2.1 Tetrahydrobiopterin Metabolism

Tetrahydrobiopterin (BH_4) is a cofactor of phenylalanine, tyrosine, tryptophan hydroxylases, and all nitric oxide synthases (Blau et al. 2001). BH_4 plays an essential role in the synthesis of L-dopa and 5-hydroxytryptophan from tyrosine and tryptophan, respectively (the initial and rate limiting step in the biosynthesis of biogenic amines) (Blau et al. 2001; Hyland 1999).

BH_4 provides electrons during reactions catalyzed by phenylalanine, tyrosine and tryptophan hydroxylases, and is oxidized to its hydroxyl compound – pterin-4α-carbinol-amine which is further converted to quininoid dihydropterin (qBH_2) by pterin-4 α-carbinolaminedehydratase (PCD; EC 4.2.1.96) and next regenerated to BH_4 by dihydrobiopterin reductase (DHPR; EC 1.6.99.7) (Blau et al. 2001).

The synthesis of BH4 involves three main reactions. The first step of the BH_4 synthesis is conversion of guanosine triphosphate (GTP) to dihydroneopterin triphosphate (NH_2P_3) by rate-limiting enzyme GTP cyclohydrolase I (GTPCH I; E 3.5.4.16). GTPCH activity is regulated by the regulatory protein GFPR and phenyloalanine concentrations. A defect in GTPCH I exists in both autosomal recessive (arGTPCH) and dominant (adGTPCH) forms. The second enzyme is 6-pyruvoyl-tetrahydropterin synthase (PTPS; EC 4.6.1.10) that converts dihydroneopterin triphosphate (NH_2P_3) to 6-pyruvoyl-tetrahydropterin (6-PTP). As the last step, sepiapterin reductase (SR; EC 1.1.1.153) reduces 6-PTP to BH_4. DHPR is an enzyme of regeneration of BH_4 from qBH_2 (Blau et al. 2001; Hyland 1999).

Defects in the biosynthesis of BH4 may occur with or without hyperphenylalaninemia (HPA). HPA is usually detected through the neonatal screening program and about 2 % of HPA detected by the newborn screening are due to disorders in BH_4 metabolism (Blau et al. 2001). DHPR, PTPS, and arGTPCH I deficiencies are common in patients with HPA and can be differentiated by pterin profile in the urine and CSF, DHPR activity in the blood, and a BH_4 loading test. PTPS deficiency increases the level of neopterin, along with normal or decreased of biopterin in the CSF and urine. DHPR usually leads to increased biopterin in the urine and CSF. ArGTPCH I causes a decrease in both neopterin and biopterin in the urine and CSF (Pearl et al. 2006; Blau et al. 2001). The autosomal dominant form of GTPCH deficiency, commonly known as the Segawa syndrome or dopa-responsive dystonia, and SR deficiency are not connected with HPA and can be detected only in CSF (Friedman et al. 2012; Kusmierska et al. 2009; Pearl et al. 2006).

The most frequent symptom of ad GTPCH deficiency is dopa-responsive dystonia with diurnal fluctuation. The major symptoms of the SR deficiency are motor delay, axial hypotonia, dystonia, and oculogyric crises with diurnal fluctuation of symptoms (Friedman et al. 2012)

There is a 'salvage pathway' for sepiapterin metabolism by its reduction by carbonyl reductase (CR) or SR to 7,8-dihydrobiopterin (BH_2) that can be further reduced to BH_4 by active dihydrofoliate reductase (DHFR). DHFR is active only in the liver, and therefore patients with SR deficiency do not have HPA and have a normal pterin profile in the urine. In the central nervous system (CNS), DHFR is not active, so that the 'salvage pathway' is not active in the brain and a deficit of SR leads to decreased concentrations of BH_4 and biogenic amine metabolites in CSF (Blau et al. 2001). SR deficiency cannot be detected by conventional methods applied for the measurement of biogenic monoamine metabolites in CSF, as it must be distinguished from all pterins in CSF, especially BH_2 and sepiapterin. The detection method is based on the finding of a marked decrease in SR activity in fibroblasts, along with SR gene (*SPR* gene) molecular analysis (Friedman et al. 2012; Kusmierska et al. 2009).

All of the enzyme defects impair BH_4 synthesis and lead to decreased concentrations of homovanillic acid (HVA) and 5-hydroxyindoloacetic acid (5-HIAA), the metabolites of dopamine and serotonin, respectively, in CNS (Blau et al. 2001; Hyland 1999).

2.2 Biogenic Amines Metabolism

Tyrosine hydroxylase (TH; EC 1.14.16.2), aromatic L-amino acid decarboxylase (AADC; EC 4.1.1.28), dopamine β-hydroxylase (DβH; EC 1.14.17.1), and monoamine oxidase (MAO; EC 1.4.3.4) deficiencies are the known deficiencies of biogenic amine metabolism (Hyland 1999).

The TH is the rate-limiting enzyme in the biosynthesis of catecholamines (dopamine, norepinephrine, and epinephrine) (Fig. 1). TH deficiency leads to decreased concentration of HVA in CSF. 5-HIAA should be normal, but sometimes, a slight decrease of 5-HIAA is observed. A low concentration of HVA without any other biochemical disturbances also can suggests an adGTPCH I deficiency. A phenylalanine loading test is helpful to differentiate these two defects. TH deficiency can be confirmed only by molecular analysis (Blau et al. 2001; Hyland 1999). The most characteristic clinical symptoms of TH deficiency are psychomotor retardation, extrapyramidal symptoms (dystonic and choreoathetotic movements, and parkinsonian symptoms), and an oculogyric crisis (Hoffmann et al. 2003).

The aromatic L-amino acid decarboxylase (AADC) enzyme is the last enzyme in the chain of dopamine and serotonin synthesis. AADC deficiency is connected with decreased concentrations of HVA and 5-HIAA and increased concentrations of 3-O-methyl-dopa (3-OMD) and 5-hydroxytryptophan (5-HT) in CSF. Additionally, in AADC deficiency vanillactic acid (VLA) and vanillpyruvic acid (VPA) occur in the urine, which may be detected by gas chromatography–mass spectrometry (GC-MS). AADC deficiency can also occur with decreased concentration of 5-methyltetrahydrofoliate (5-MTHF) (Clayton 2006). The characteristic clinical symptoms of AADC deficiency are similar to those of TH deficiency above outlined. In addition, there may be truncal hypotonia, limb hypertonia, body temperature instability, and hypoglycemia (Ide et al. 2009; Swoboda et al. 2003). A mutation in *SLC18A2* gene, which encodes VMAT2, results in a phenotype which overlaps with monoamine disorders (Rilstone et al. 2012).

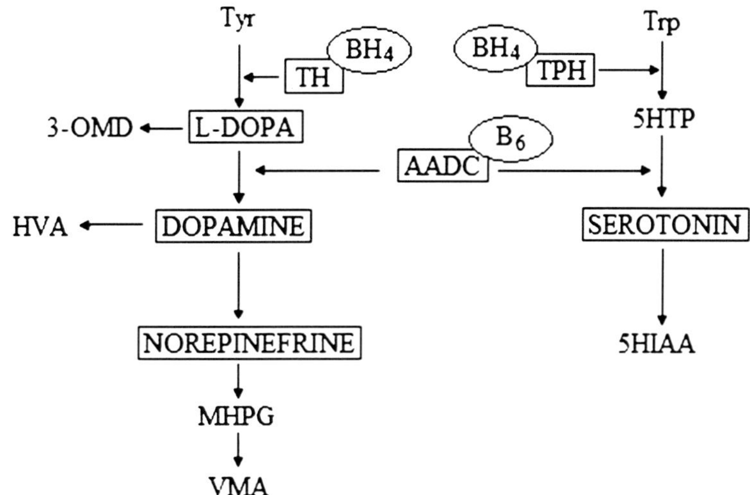

Fig. 1 Schematic presentation of biogenic amine metabolism. *Tyr* tyrosine, *Trp* tryptophan, *TH* tyrosine hydroxylase, *TPH*, tryptophan hydroxylase, *BH4* tetrahydrobiopterin, *5HTP* 5-hydroxytryptophan. *5HT* 5- hydroxytryptamine, *AADC* aromatic L-amino acid decarboxylase, *5HIAA* 5-hydroxyindoleacetic acid, *MHPG*, 3-methoxy-4-hydroxyphenylglycol, *HVA* homovanillic acid, *VMA*, vanillylmandelic acid

2.3 Foliate Metabolism

Folic acid is not a neurotransmitter, but it is closely linked to neurotransmitters' metabolism (Hyland et al. 2010). Foliate is a main cofactor in DNA biosynthesis, necessary for the methylation process. It is a critical determinant of the embryonic central nervous system development and participates in the synthesis of purines, pyrimidines, and in the metabolism of serine, histidine, methionine, and glycine. Folic acid also affects the activity of some enzymes, such as DHPR, AADC, and 3-phosphoglycerate dehydrogenase. 5-methyl-tetrahydrofolate (5-MTHF) is decreased in CSF of patients with a low folic acid concentration. Folic acid has three separate transporter systems (Hyland et al. 2010). Deficiency of the primary transporter may lead to cerebral foliate deficiency (CFD) and decreased 5-MTHF concentration in CSF. A defective secondary system may be confirmed by increases in HVA and 3-OMD in CSF.

2.4 γ-Aminobutyric Acid (GABA) Metabolism

GABA is a major inhibitory neurotransmitter of the brain and is formed from glutamic acid by the B_6-dependent enzyme – glutamate decarboxylase (GAD; EC4.1.1.15). GABA is transported into synaptic vesicles by the vesicular GABA transporter (VGAT). GABA is converted to succinic semialdehyde (SSA) in the reaction catalyzed by GABA transaminase (GABA T; EC 2.6.1.19) (Jakobs et al. 1981). Then, succinic semialdehyde is oxidized by succinic semialdehyde dehydrogenase (SSADH; EC 1.2.1.24) to succinate that is successively used in Krebs cycle. Two defects of the GABA metabolism have been reported: GABA-T deficiency and SSADH deficiency (Jaeken et al. 1984; Jakobs et al. 1981).

The best known neurotransmitter disorder is SSADH deficiency. The main symptoms of SSADH deficiency are a developmental delay and intellectual disability, behavioral problems, motor dysfunction (ataxia), and epilepsy (Kim et al. 2011). GABA-T deficiency is a rare disorder of GABA catabolism, with a severe psychomotor retardation and recurrent episodic lethargy accompanied by intractable seizures (Tsuji et al. 2010).

2.5 Pyridoxine-Dependent Enzymes

Vitamin B_6 is present in the organism as six vitamers: pyridoxine (pyridoxol), pyridoxamine, pyridoxal, and their 5′-phosphorylated esters. The biologically active form of pyridoxine (cofactor activity) is pyridoxine-5-phosphate (PLP) which is a cofactor for approximately 100 enzymes. Pyridoxine dependency is caused by the binding of PLP to abnormal metabolites formed due to the deficiency of alpha-aminoadipic semialdehyde dehydrogenase. AADC also depends on the PLP; thus a PLP disorder affects the biogenic amine metabolite profile in CSF in a way similar to the primary low AADC activity. PLP deficiency occurs with increased glycine concentration in CNS, causing neurological disturbances. Lack of PLP is the most commonly caused by decreased pyridoxine-5-phosphate oxidase (PNPO) activity. PNPO deficiency leads to severe neurological sequelae. A typical clinical presentation is characterized by perinatal onset, epileptic encephalopathy, refractoriness to antiepileptic drugs, and responsiveness to PLP treatment (Rahman et al. 2012; Clayton 2006; Pearl et al. 2006; Jaeken et al. 1984; Jakobs et al. 1981).

2.6 Glycine Encephalopathy

Glycine is an inhibitory neurotransmitter in the spinal cord and acts as a co-agonist of the glutamatergic NMDA receptor. An increased glycine concentration in CSF is usually caused by a defect of the glycine cleavage system. The system has four components: a P-protein containing a PLP-dependent glycine decarboxylase; H-protein; T-protein, which is tetrahydrofolate-dependent, and L-protein, which is a lipoamide dehydrogenase moiety of the pyruvate dehydrogenase complex (Rahman et al. 2012; Pearl et al. 2006). The P-protein defect is associated with the neonatal-onset forms, while H- and T-protein defects are associated with the later-onset forms. A classic neonatal phenotype includes *in utero*-onset seizures. In the neonatal form, clinical findings include neonatal encephalopathy, hypotonia, myoclonias, and apnea. A later childhood variant includes mild mental retardation, delirium, chorea, and vertical gaze palsies, while a late onset pattern in adults is characterized by progressive spastic diplegia and optic atrophy. Defects of the glycine cleavage system are detected by a CSF to plasma glycine ratio of more than 0.08. The measurement of glycine cleavage activity in the liver and cultured lymphoblasts, or genetic identification are available for accurate diagnosis (Rahman et al. 2012; Pearl et al. 2006).

3 Receptors and Protein Transporters

Inborn errors of neurotransmitters involve defects in biogenic amines, GABA, glycine, and in the metabolism of the cofactors BH_4 and PLP. Likewise, pathologic alterations of proteins involved in synaptic neurotransmission may cause neurological and psychiatric symptoms and abnormalities in the CSF content (Bagale et al. 2011; Duarte et al. 2011). Therefore, in patients with low CSF levels of HVA and 5-HIAA, unrelated to an enzymatic defect, further investigation is warranted to discern alternative etiologies, such as receptor or transporter defects. This may be done by positron emission tomography (PET) and single-photon emission computed tomography (SPECT), both representing unique techniques for the assessment of *in vivo* dopamine receptor distribution in humans. PET neuroimaging can be also used to evaluate other types of vesicular presynaptic transport systems (Lee et al. 2009; Breit et al. 2006; Buu 1989). Neuroimaging may be a useful method for patients with tyrosine hydroxylase or AADC deficiency. The activity of AADC can be detected *in vivo* using ^{18}fluoro-dopa-PET (^{18}F-dopa-PET) (Martin et al. 1989).

Monoamine and some amino acid NTs transporters belong to the solute carrier 6 (SLC6) family, e.g., the transporter for GABA (Bagale et al. 2011; Duarte et al. 2011). The function of these transporters depends on the transmembrane Na^+ and Cl^- gradient. The dopamine transporter (DAT) plays a key role in the regulation of dopaminergic signaling in the brain by moving dopamine from the synaptic cleft into the cytoplasm. The vesicular monoamine transporter-2 (VMAT2) is responsible for the loading of monoamine into the synaptic vesicles. VMAT2 and VGAT vesicular transporters are intracellular proteins. In turn, D2R is a receptor that belongs to the G-protein receptor family. The serotonin transporter (SERT), also belonging to the SLC6 family, is a transporter responsible for the active re-uptake of serotonin from the synapse.

4 Clinical Presentation of Inherited Neurotransmitter Disorders

Patients with inherited neurotransmitter disorders present various clinical symptoms and signs (Friedman et al. 2012; Bagale et al. 2011; Kim et al. 2011; Tsuji et al. 2010; Kusmierska et al. 2009; Clayton 2006; Pearl et al. 2006; Pearl et al. 2005a, b; Swoboda et al. 2003; Hoffman et al. 2003; Blau et al. 2001; Hyland 1999).

4.1 Newborns

- Abnormal muscle tone in the form of significant hypotonia or hypertonia;
- Seizures refractory to anticonvulsants;
- Positive screening test of phenylketonuria with abnormal pterin profile in CSF (BH_4 deficiency);
- Additional clinical features include jitteriness, hypothermia, neonatal dystonia, restlessness, irritability, and emesis preceding seizures.

4.2 Children Older than 1 Month

- Extrapyramidal and pyramidal-extrapyramidal syndromes of unknown etiology, progressive nature of disease, daily fluctuations, and normal brain MRI;
- Hypotonic child;
- Hypertonic child (stiff baby);
- Regulatory disorders in infants not being ameliorated with age, accompanied by psychomotor;
- Retardation and impaired locomotor patterns;
- Drug-resistant epilepsy;
- Other symptoms implying neurotransmitter disorders; oculogyric crisis;
- Progressive encephalopathy of unknown etiology.

Every patient should undergo a structured interview and a detailed examination including neurological, biochemical, and imaging. Lumbar puncture must be performed to obtain the material for clinical diagnosis. CSF should be collected in five fractions following a strictly defined protocol and samples are to be stored at −80 °C until analysis. Metabolites of biogenic amines can be examined with high performance liquid chromatography (HPLC) with coulometric detection. HPLC with fluorimetric detection can be used to analyze pterin profile. 5-MTHF can be measured by HPLC with a coulometric detector (Kusmierska et al. 2009; Blau et al. 2001). In case of an abnormal result, further tests, such as oral phenylalanine loading test and amino acid and organic acid analysis in the urine by GC-MS, should be performed.

5 Neurotransmitter System as a Whole

Efficient communication in the brain requires a precise control of neurotransmitter action at specific molecular targets. Each neuron is connected with numerous other neurons, receiving signals by various NTs. Moreover, it has been shown that a single presynaptic terminal could release several NTs (co-transmission). Some NTs can be co-released to provide appropriate balance of the signal and to enable more complex communication. The knowledge about co-release of NTs and co-transmission in humans is still limited. Only have a limited number of studies directly examined more than one neurotransmitter system at a time and explored mutual associations of these systems in neurometabolic disorders in humans. Disorders that affect the metabolism of NTs are complex and interactions between different metabolic pathways can lead to severely compromised neurological function (Duarte et al. 2011). Inherited NTs disorders attributed to a primary disturbance in NTs metabolism or transport are by far poorly understood. Deregulation of several different neurotransmitter systems is always implicated in the etiology of neurological disorders. It is evident that dysfunction of large-scale brain networks rather than differences in single regions underlie psychiatric disorders. Little is known about the correlation between abnormal metabolism of biogenic amines (primary or secondary) and abnormalities of the metabolism of other neurotransmitters. Furthermore, the relationship between metabolic pathways of a large number of newly characterized neurotransmitters is poorly characterized. A concomitant measurement of different NTs, their precursors, metabolites, receptors, and cofactors at the same time-point in CSF may provide a more accurate picture of brain function and help understand the mechanisms of cerebral dysfunctions and create novel therapeutic options. The rationale for using the CSF as a source of various metabolites to study neurotransmitter interactions and their receptor or protein transporter connections seems to have clinical relevance and may yield information otherwise unavailable from a single neurotransmitter study.

Conflicts of Interest The authors declare no conflicts of interest in relation to this article.

References

Anderson CM, Kidd PD, Eskandari S (2010) GATMD: gamma-aminobutyric acid transporter mutagenesis database. Database (Oxford): baq028

Bagale SM, Brown AS, Carballosa Gonzales MM, Vitores A, Micotto TL, Saleesh Kumar NS, Hentall ID, Wilson JN (2011) Fluorescent reporters of monoamine transporter distribution and function. Bioorg Med Chem Lett 21:7387–7391

Blau N, Thöny B, Cotton RGH, Hyland K (2001) Disorders of tetrahydrobiopterin and related biogenic amines. In: Scriver CR, Beaudet AL, Sly WS, Valle D, Childs B, Vogelstein B (eds) The metabolic and molecular bases of inherited disease. McGraw-Hill, New York, pp 1725–1776

Breit S, Reimold M, Reischl G, Klockgether T, Wullner U (2006) [^{11}C]d-threo-methylphenidate PET in patients with Parkinson's disease and essential tremor. J Neural Transm 113:187–193

Buu NT (1989) Vesicular accumulation of dopamine following l-dopa administration. Biochem Pharmacol 38:1787–1792

Clayton PT (2006) B6-responsive disorders: a model of vitamin dependency. J Inherit Metab Dis 29:317–326

Duarte ST, Ortez C, Pérez A, Artuch R, García-Cazorla A (2011) Analysis of synaptic proteins in the cerebrospinal fluid as a new tool in the study of inborn errors of neurotransmission. J Inherit Metab Dis 34:523–528

Friedman J, Roze E, Abdenur JE, Chang R, Gasperini S, Saletti V, Wali GM, Eiroa H, Neville B, Felice A, Parascandalo R, Zafeiriou DI, Arrabal-Fernandez L, Dill P, Eichler FS, Echenne B, Gutierrez-Solana LG, Hoffmann GF, Hyland K, Kusmierska K, Tijssen MA, Lutz T, Mazzuca M, Penzien J, Poll-The BT, Sykut-Cegielska J, Szymanska K, Thöny B, Blau N (2012) Sepiapterin reductase deficiency: a treatable mimic of cerebral palsy. Ann Neurol 71:520–530

Hoffmann GF, Assmann B, Brautigam C (2003) Tyrosine hydroxylase deficiency causes progressive encephalopathy and dopa-nonresponsive dystonia. Ann Neurol 54(Suppl 6):S56–S65

Hyland K (1999) Neurochemistry and defects of biogenic amine neurotransmitter metabolism. J Inherit Metab Dis 22:353–363

Hyland K, Shoffner J, Heales SJ (2010) Cerebral folate deficiency. J Inherit Metab Dis 33:563–570

Ide S, Sasaki M, Kato M, Shiihara T, Kinoshita S, Takahashi JY, Goto Y (2009) Abnormal glucose metabolism in aromatic L-amino acid decarboxylase deficiency. Brain Dev 32:506–510

Jaeken J, Casaer P, De Cock P, Corbeel L, Eeckels R, Eggermont E, Schechter PJ, Brucher JM (1984) Gamma-aminobutyric acid-transaminase deficiency: a newly recognized inborn error of neurotransmitter metabolism. Neuropediatrics 15:165–169

Jakobs C, Bojasch M, Monch E, Rating D, Siemes H, Hanefeld F (1981) Urinary excretion of γ-hydroxybutyric acid in a patient with neurological abnormalities. The probability of a new inborn error of metabolism. Clin Chim Acta 111:169–178

Kim KJ, Pearl PL, Jensen K, Snead OC, Malaspina P, Jakobs C, Gibson KM (2011) Succinic semialdehyde dehydrogenase: biochemical-molecular-clinical disease mechanisms, redox regulation, and functional significance. Antioxid Redox Signal 15:691–718

Kusmierska K, Jansen EE, Jakobs C, Szymanska K, Malunowicz E, Meilei D, Thony B, Blau N, Tryfon J, Rokicki D, Pronicka E, Sykut-Cegielska J (2009) Sepiapterin reductase deficiency in a 2-year-old girl with incomplete response to treatment during short-term follow-up. J Inherit Metab Dis 32:S5–S10

Lee WT, Weng WC, Peng SF, Tzen KY (2009) Neuroimaging findings in children with pediatric neurotransmitter diseases. J Inherit Metab Dis 32:361–371

Lin Z, Canales JJ, Bjorgvinsson T, Thomsen M, Qu H, Liu QR, Torres GE, Caine CB (2011) Monoamine transporters: vulnerable and vital doorkeepers. Prog Mol Biol Transl Sci 98:1–46

Martin WR, Palmer MR, Patlak CS, Calne DB (1989) Nigrostriatal function in humans studied with positron emission tomography. Ann Neurol 26:535–542

Pearl PL, Bennet HD, Khademian Z (2005a) Seizures and metabolic disease. Curr Neurol Neurosci Rep 5:127–133

Pearl PL, Capp PK, Novotny EJ, Gibson KM (2005b) Inherited disorders of neurotransmitters in children and adults. Clin Biochem 38:1051–1058

Pearl PL, Hartka TR, Taylor J (2006) Diagnosis and treatment of neurotransmitter disorders. Curr Treat Options Neurol 8:441–450

Rahman S, Footitt EJ, Varadkar S, Clayton PT (2012) Inborn errors of metabolism causing epilepsy. Dev Med Child Neurol 55:23–36

Rilstone JJ, Alkhater RA, Minassian BA (2012) Brain dopamine-serotonin vesicular transport disease and its treatment. N Engl J Med 368:543–550

Swoboda KJ, Saul JP, McKenna CE, Speller NB, Hyland K (2003) Aromatic L-amino acid decarboxylase deficiency: overview of clinical features and outcomes. Ann Neurol 54:S49–S55

Tsuji M, Aida N, Obata T, Tomiyasu M, Furuya N, Kurosawa K, Errami A, Gibson KM, Salomons GS, Jakobs C, Osaka H (2010) A new case of GABA transaminase deficiency facilitated by proton MR spectroscopy. J Inherit Metab Dis 33:85–90

Advs Exp. Medicine, Biology - Neuroscience and Respiration (2015) 6: 9–17
DOI 10.1007/5584_2014 72

Published online: 14 October 2014

Inhibition of Peripheral Dopamine Metabolism and the Ventilatory Response to Hypoxia in the Rat

Monika Bialkowska, Dominika Zajac, Andrea Mazzatenta, Camillo Di Giulio, and Mieczyslaw Pokorski

Abstract

Dopamine (DA) is a putative neurotransmitter in the carotid body engaged in the generation of the hypoxic ventilatory response (HVR). However, the action of endogenous DA is unsettled. This study seeks to determine the ventilatory effects of increased availability of endogenous DA caused by inhibition of DA enzymatic breakdown. The peripheral inhibitor of MAO – debrisoquine, or COMT – entacapone, or both combined were injected to conscious rats. Ventilation and its responses to acute 8 % O_2 in N_2 were investigated in a whole body plethysmograph. We found that inhibition of MAO augmented the hyperventilatory response to hypoxia. Inhibition of COMT failed to influence the hypoxic response. However, simultaneous inhibition of both enzymes, the case in which endogenous availability of DA should increase the most, reversed the hypoxic augmentation of ventilation induced by MAO-inhibition. The inference is that when MAO alone is blocked, COMT takes over DA degradation in a compensatory way, which lowers the availability of DA, resulting in a higher intensity of the HVR. We conclude that MAO is the enzyme

M. Bialkowska and D. Zajac
Department of Respiratory Research, Medical Research Center, Polish Academy of Sciences, Warsaw, Poland

A. Mazzatenta
Neuroscience Institute, C.N.R., Pisa, Italy

Department of Neuroscience and Imaging, "G. d'Annunzio" University of Chieti-Pescara, Chieti, Italy

C. Di Giulio
Department of Neuroscience and Imaging, "G. d'Annunzio" University of Chieti-Pescara, Chieti, Italy

M. Pokorski (✉)
Department of Nursing, Public Higher Medical Professional School in Opole, 68 Katowicka St., 45-060 Opole, Poland

Institute of Psychology, Opole University, 1 Plac Staszica, 45-052 Opole, Poland
e-mail: m_pokorski@hotmail.com

predominantly engaged in the chemoventilatory effects of DA. Furthermore, the findings imply that endogenous DA is inhibitory, rather than stimulatory, for hypoxic ventilation.

Keywords
Carotid body • Catechol-O-methyltransferase • Dopaminergic system • Endogenous dopamine • Hypoxic ventilatory response • Minute ventilation • Monoaminooxidase

1 Introduction

Dopamine (DA) is present in a substantial amount and is considered a putative neurotransmitter in the carotid body (Hellström 1977), a paired sensory organ of neural crest origin whose chemoreceptor cells generate the excitatory response to chemical stimuli, most notably to reductions in partial oxygen pressure in arterial blood (PaO_2). DA is released from carotid chemoreceptor cells in proportion to the strength of the hypoxic stimulus and binds to D2 receptors on the plasma membrane to trigger the cellular transduction cascade, ending up in increased discharge rate in the sinus nerve endings apposing chemoreceptors (Gonzalez et al. 1994). The carotid body discharge is then relayed to the brain stem respiratory areas to evoke a hyperventilatory response (Faff et al. 1999).

Exogenous DA and D2 receptor antagonists have been extensively used as pharmacological tools, taking advantage of the incapability of the hydrophilic DA to cross the blood-brain barrier, which enables to study the peripheral DA-mediated effects without vagueness. Nevertheless, studies failed to determine the exact role of DA in ventilatory regulation and the issue remains contentious. DA seems to have a well established inhibitory role for ventilation and its responses to acute hypoxia in the majority of species, such as the cat (Llados and Zapata 1978), the rabbit (Matsumoto et al. 1980), or the rat (Bee and Pallot 1995; Monteiro et al. 2011), but not in the dog where it has a stimulatory effect (Black et al. 1972). However, the notion of DA-mediated ventilatory inhibition has found support in the action of domperidone, a peripheral D2 antagonist, which increases ventilation (Bee and Pallot 1995), although in some reports the increase has only been found after birth and was lost with maturation (Tomares et al. 1994). The issue is further confounded by the postulate of the existence of the low affinity excitatory post-synaptic carotid body dopamine D2 receptor responding to high doses of DA, as opposed to the inhibitory effects on the hypoxic ventilator responses exerted by low doses of DA through high affinity D2 receptors in animals and man (Gonzales et al. 1994; Ward and Bellville 1982). The probable presence of two subtypes of D2 receptors, differing in affinity to DA, makes an understanding of the action of endogenous DA, released in a small concentration, unclear.

On the premise that DA is basically inhibitory for ventilation as mentioned above we made the hypothesis that the stimulatory ventilatory response to hypoxia might be attenuated by the inhibition of DA breakdown, and thus enhancement of its endogenous availability for the D2 receptors within the carotid body. There are reports showing that blockade of MAO and COMT effectively increases the dopamine content in a tissue (Wang et al. 2001). Therefore, we addressed this issue in conscious rats by using specific peripheral antagonists of the two enzymes responsible for DA degradation: monoaminooxidase (MAO) and catechol-O-methyltransferase (COMT). We established that MAO is the predominating enzyme engaged in the chemosensory effects of DA degradation in the carotid body. Although inhibition of MAO augmented the hyperventilatory response to hypoxia, the augmentation was abrogated by the simultaneous addition of COMT; the situation when the accumulation of endogenous DA should be the most. The inference is that endogenous DA is inhibitory for ventilation. The abrogation of MAO-induced ventilatory stimulation, when both COMT and MAO inhibitors were used, could be due to COMT taking over DA degradation when MAO alone is blocked; a phenomenon reported in the literature (Trendelenburg 1984).

2 Methods

2.1 Animals and Instrumentation

The study was approved by the IV Local Ethics Committee for Animal Experiments in Warsaw, Poland (Permit Number: 29/2010) and was conducted in accord with the guiding principles for the European Convention for the Protection of Vertebrate Animals used for Experimental and other Scientific Purposes (Council of Europe No 123, Strasbourg 1985). A total of 26 adult male, conscious Wistar rats, weighing 299.9 ± 3.0 (SE) g were used. All rats were kept on a 12 h light-dark cycle, temperature of 21 ± 2 °C, humidity of 50–60 %, and were fed with a standard animal chow and had water *ad libitum*. The animals were divided into the following 4 groups: control group (n = 5) that received the vehicle – 0.3 ml dimethylsulfoxide (DMSO), the debrisoquine group (n = 7) that received 40 mg/kg of the peripheral MAO inhibitor debrisoquine, the entacapone group (n = 7) that received 30 mg · kg^{-1} of the peripheral COMT inhibitor entacapone, and finally the last group (n = 7) received both inhibitors in the above-mentioned doses. All drugs were administered intraperitoneally in a 0.3 ml volume; entacapone being dissolved in DMSO and debrisoquine in physiological saline, and the injections of both inhibitors were made 2 min apart symmetrically into either side of the peritoneal cavity. Entacapone was generously provided by Orion Corporation – Orion Pharma (Espoo, Finland) and debrisoquine was purchased from Sigma-Aldrich (St. Louis, MO).

Lung ventilation and its responses to acute hypoxia were measured in a whole body rodent plethysmograph (model PLY3223; Buxco Electronix Inc., Wilmington, NC). Each unrestricted rat was placed in the recording chamber. Chamber temperature was maintained constant at 21 °C throughout the experiment. Bias flow at a rate of 2.5 l · min^{-1} between the hypoxic tests was used, via a flow pump reservoir system (PLY1020, Buxco Electronics), for removing CO_2 build-up from the chamber. Pressure difference between the experimental and reference chamber was measured with a differential pressure transducer. The pressure signal was amplified and then integrated by data analysis software (Biosystem XA for Windows SFT3410 v. 2.9; Buxco Electronics).

2.2 Ventilatory Measurements

Volume (V_T) and frequency (f) components of lung ventilation were measured breath-by-breath were processed to yield instantaneous minute ventilation (V_E, ml · min^{-1}, BTPS). Data were further processed off-line to show the 30 s time points of the 180 s course of the hypoxic ventilatory response. Each point constituted an average of a 10 s bin of a given variable preceding the 30 s time mark.

2.3 Study Protocol

Each animal was used once. At the beginning of the experiment, the animal was allowed to accustom to the chamber in ambient air for about 15 min. Then, the gas in the chamber was switched to 8 % O_2 in N_2. The equilibrium of a gas mixture in the chamber was achieved within 40 s, after which a 3-min hypoxic poikilocapnic recording started. During the recovery period in room air, the DMSO vehicle or the enzymes' inhibitors were injected according to the scheme and doses above outlined. The hypoxic tests were repeated after 30 and 60 min from the injection in like manner.

2.4 Blood Pressure

Blood pressure was measured noninvasively with a CODA tail-cuff blood pressure system using a volume pressure recording (VPR) sensor technology and software that enables to continuously monitor data in real-time (Kent Scientific, Torrington, CT, USA).

2.5 Data Evaluation

Data were presented as means ± SE. Ventilatory variables were normalized to weight in kg. The data turned out to be normally distributed; checked with the Shapiro-Wilk test. Statistical elaboration took into account three main points of interest characterizing V_E changes along the hypoxic time course: prehypoxic baseline, peak hypoxic increase, and hypoxic depressant nadir. One-way ANOVA was applied to compare V_E at these time points across each hypoxic profile and also across the corresponding time points of the three experimental conditions: untreated control, 30 min, and 60 min after a given pharmacological intervention. If significant, the source of differences was further evaluated with the *post-hoc* Scheffe test. Statistical significance was defined as $P < 0.05$.

3 Results

All hypoxic responses recorded in the present study had a classical biphasic stimulatory/inhibitory character. V_E peaked at about 30 s from the start of hypoxia, which was followed by a gradual roll-off reaching nadir within 150–180 s. In this depressant phase, V_E decreased by about 15–20 % of the peak increase, remaining significantly above the baseline prehypoxic level.

The results show that debrisoquine, 30 min after the injection, significantly enhanced both the resting baseline ventilation and the ventilatory response to 8 % hypoxia along its recorded course. The enhancement was most pronounced at peak response where V_E increased from 1.52 ± 0.82 before to $2.33 \pm 0.30\ \mathrm{l \cdot min^{-1} \cdot kg^{-1}}$ after debrisoquine; $P < 0.05$ (Fig. 1A). However, this effect failed to be long-lived. The ventilatory response to hypoxia repeated 60 min after debrisoquine injection showed that the peak V_E regressed to $1.66 \pm 0.12\ \mathrm{l \cdot min^{-1} \cdot kg^{-1}}$; a downfall also seen at the following time marks of the hypoxic profile, nearly to the level present in the control untreated condition. The increases in hypoxic ventilation after debrisoquine were achieved due to contributions of both frequency and tidal ventilatory

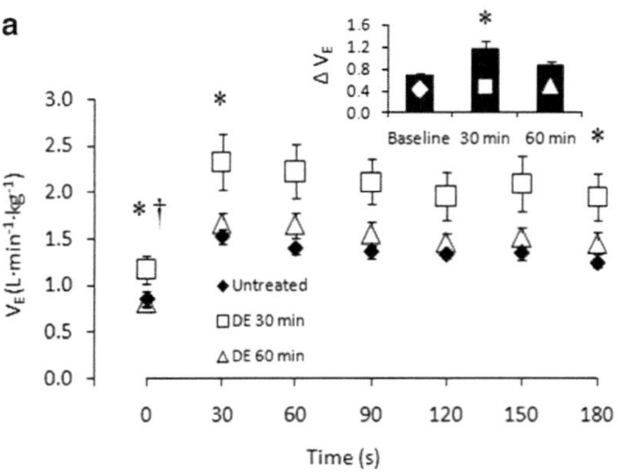

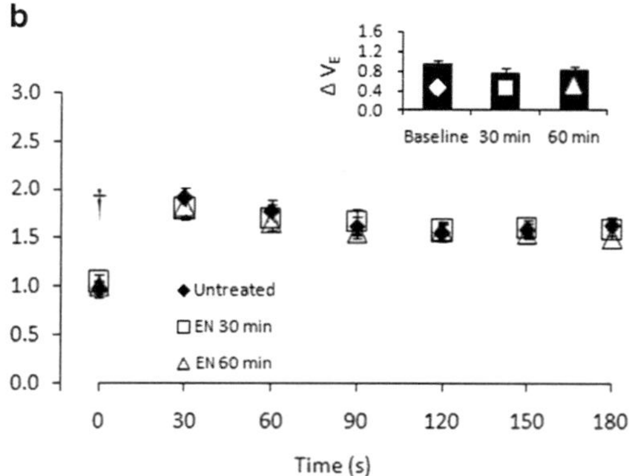

Fig. 1 Hypoxic ventilatory responses in the untreated control condition, and then 30 and 60 min after i.p. administration of debrisoquine (*A*) and entacapone (*B*). *p < 0.05 for the differences at the corresponding time marks (*vertical*) as follows: at 0 s mark – between 30 and 60 min post-debrisoquine V_E *vs.* untreated prehypoxic baseline V_E; at 30 s mark – between 30 min post-debrisoquine peak hypoxic V_E *vs.* untreated peak V_E, and at 180 s mark – between post-debrisoquine hypoxic V_E nadir *vs.* untreated V_E nadir; †p < 0.05 for the differences along the sequential time marks (*horizontal*) of the hypoxic courses as follows: prehypoxic untreated V_E *vs.* the peak and nadir V_E in each experimental condition in both panels. The insets show the augmentation of V_E from the normoxic baseline to hypoxic peak 30 and 60 min after debrisoquine (*Panel A*; *significantly higher 30 min after debrisoquine, P < 0.05) (*Panel B* – insignificant differences)

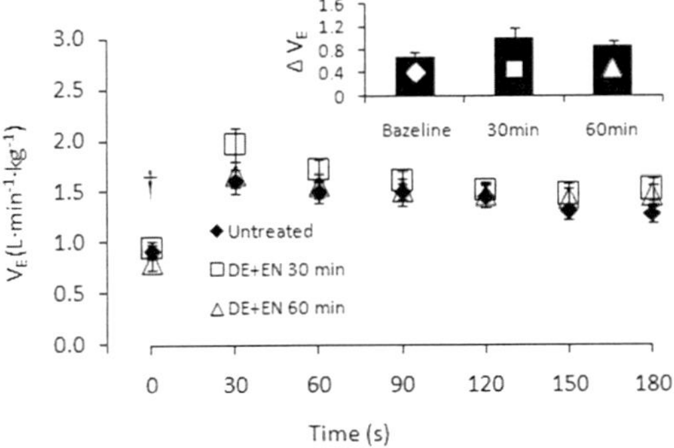

Fig. 2 Hypoxic ventilatory responses in the untreated control condition, and then 30 and 60 min after debrisoquine and entacapone administered together. There were no significant differences noted at the corresponding time marks (*vertical*) of the three experimental conditions. †Prehypoxic untreated V_E (0 s time mark) significantly lower than peak (30 s time mark) and nadir (180 s time mark) V_E in each of the three experimental conditions; $p < 0.05$. The *inset* shows the augmentation of V_E from normoxic baseline to hypoxic peak 30 and 60 min post-debrisoquine+entacapone; the differences being insignificant

components. Peak breathing frequency amounted to 139 ± 10 breaths·min^{-1} in the untreated condition and changed to 168 ± 12 and 145 ± 10 breaths · min^{-1} at 30 and 60 min post-debrisoquine. The corresponding values for tidal component were 12.0 ± 1.6, 14.0 ± 1.6, and 11.0 ± 0.6 ml · kg^{-1}. Although the joint action of both components led to significant changes in V_E, changes in either component turned out to be insignificant ($P > 0.05$, one-way ANOVA).

Entacapone, on the other hand, failed to appreciably affect the hypoxic ventilatory response either 30 or 60 min after injection (Fig. 1B). Likewise, entacapone combined with debrisoquine failed to affect the course of the hypoxic ventilatory response in a significant way, although there was a tendency remaining for a higher peak hypoxic V_E compared with the untreated condition, which was gone 60 min after injection ($P > 0.05$) (Fig. 2). Thus, the ventilatory augmentation evoked by MAO inhibition alone was gone when the COMT inhibitor was simultaneously used.

Hypoxic ventilatory responsiveness remained grossly unchanged after DMSO-vehicle injection, compared with that before DMSO, during the 60 min time span recorded, although DMSO showed a slight tendency to dampen both baseline and hypoxic ventilatory levels (Table 1).

Neither debrisoquine nor entacapone caused any meaningful changes in arterial blood pressure recorded up to 90 min after the injection (Table 2).

4 Discussion

This investigation demonstrates that MAO-mediated oxidation is the major pathway of dopamine degradation in the carotid body as judged by changes in the hypoxic chemoreflex in response to pharmacological blockade of the enzyme. The predominant role of MAO in DA metabolism is in line with the prevailing presence of MAO in many a cell type, such as peripheral neurons, glial cells, and others (Weyler et al. 1990; Hovevey-Sion et al. 1989). Separate inhibition of the COMT pathway remained without any ventilatory effects, showing no basic role of COMT in DA degradation.

The augmentation of the hyperventilatory response to hypoxia after the peripheral MAO inhibition seemingly ran counter to our working presumption that MAO inhibition, by slowing down DA metabolism and increasing the availability of DA at its functional receptor sites in carotid body chemoreceptors, ought to bring up the inhibitory character of DA regarding the ventilatory regulation. However, simultaneous inhibition of both COMT and MAO, the condition in which endogenous DA should accumulate the most, reversed the hypoxic augmentation of ventilation induced by MAO-inhibition. These results imply the biological plausibility that when MAO alone is blocked, COMT takes over DA degradation in a compensatory way, which lowers the availability of DA, resulting in a higher intensity of the hypoxic response. The corollary is endogenous DA is in fact inhibitory for ventilation. The ability of COMT to compensate for the lost function of MAO, but not the other way around, has been reported in the rat heart (Trendelenburg 1984).

Table 1 Responses of minute ventilation (V_E) to 8 % hypoxia before and after administration of DMSO-vehicle

			After DMSO					
Before DMSO			30 min			60 min		
Baseline V_E	Hypoxic peak V_E	Hypoxic nadir V_E	Baseline V_E	Hypoxic peak V_E	Hypoxic nadir V_E	Baseline V_E	Hypoxic peak V_E	Hypoxic nadir V_E
1.08 ± 0.94	1.74 ± 0.16	1.39 ± 0.72	1.02 ± 0.17	1.52 ± 0.17	1.32 ± 0.13	0.93 ± 0.64	1.56 ± 0.13	1.37 ± 0.95

Values are means ± SE; $l \cdot min^{-1} \cdot kg^{-1}$. Peak hypoxic responses were noted 30 s after the initiation of hypoxic breathing. No significant differences were noted in the hypoxic ventilatory responses before and after DMSO

Table 2 Influence of entacapone and debrisoquine on arterial blood pressure

	Mean arterial blood pressure (mmHg)	
Time (min)	Debrisoquine	Entacapone
0	119 ± 7	116 ± 5
30	105 ± 13	128 ± 7
90	123 ± 16	133 ± 8

Values are means ± SE. Insignificant changes

The role of DA in carotid body function is a highly contentious issue. DA is present in the carotid body in high amounts in chemoreceptor cells and shapes their sensory responses, being released in proportion to the hypoxic stimulus strength (Gonzalez et al. 1994). DA infusion inhibits carotid body responses to hypoxia and DA D_2 receptor blockade increases these responses in most species (Monteiro et al. 2009; Chow et al. 1986; Eyzaguirre and Zapata 1984). The issue is further confounded by the apparent discrepancy in translation of the sensory carotid body discharge into ventilatory outcome. Smatresk et al. (1983) have reported that a non-specific D2 antagonist, haloperidol, injected intravenously in a dose of 1 mg/kg in anesthetized cats, increases the carotid sensory discharge in the sinus nerve, but attenuates the ventilatory response to hypoxia. The investigators concluded that haloperidol, which penetrates into the brain, blocks the central integration of peripheral chemoreceptor input. There is some supportive evidence to this end showing no effect of haloperidol on ventilation during normoxia or hypoxia (Bainbridge and Heistad 1980), or a depressant effect on the response to hypercapnia (Lundberg et al. 1979).

DA infusion in some species, such as goats or dogs, causes initially a burst of excitatory carotid sensory discharge, particularly observed after high doses of DA, later followed by depression (Bisgard et al. 1979). Since ventilation follows carotid body discharge rate, DA can cause stimulation or inhibition of the responses to natural chemical stimuli. To reconcile these divergent effects of DA on hypoxic ventilation, Gonzalez et al. (1994) have proposed the existence of a dichotomous D2 receptor population located at the carotid chemoreceptor cell/sinus nerve ending synapses consisting of two opposed classes: low affinity post-synaptic receptors at which DA would act as an excitatory neurotransmitter and high affinity D_2 autoreceptors on chemoreceptor cells at which DA would be inhibitory, and which would regulate DA release by these cells. It follows that high doses of DA would stimulate sinus nerve activity, and consequently ventilation, through low affinity receptors and *vice versa* low doses of DA would inhibit sinus nerve activity through high affinity autoreceptors. Although we did not measure DA content in the carotid body, which would require alternative study design, the possibility of a direct facilitatory effect on ventilation of DA seems unlikely in face of the disappearance of ventilatory augmentation when both MAO and COMT were blocked simultaneously, the condition which favors DA accumulation. Our findings are in rapport with the concept that low concentrations of DA at the carotid body would act to dampen ventilation through high affinity D2 receptors.

The reversion of MAO-mediated ventilatory augmentation when both MAO and COMT were blocked also makes a plausible role of noradrenaline unlikely in the effects observed. A slowdown of DA metabolism could increase the content of noradrenaline that is formed by dopamine β-hydroxylase, and then is metabolized by both MAO and COMT. Noradrenaline, in contrast to DA, is present mostly in sympathetic nerve fibers reaching the carotid body from the superior cervical ganglion and participates in vascular and blood flow regulation (Hanbauer and Hellström 1978) rather than in the inherent chemotransduction mechanisms. There is less noradrenaline than DA in the carotid body; the ratio being about 1:5 (Vicario et al. 2000). Nonetheless, hypoxia stimulates sympathetic nerve activity and release of noradrenaline in the carotid body and noradrenaline increases the chemosensory discharge rate and ventilation (Somers et al. 1989; Joels and White 1968). Chemoreceptor cells function also is sensitive to arterial blood pressure changes (Lahiri et al. 1980) that could be induced by the accumulation of catecholamine. Lack of blood pressure changes in the present study, when both enzymes were blocked, makes the involvement of sympathetic background in the results obtained unlikely. Others studies have also reported that simultaneous blockade of MAO and COMT does not influence hemodynamics or concentrations of unconjugated norepinephrine in plasma (Illi et al. 1996).

In conclusion, the present findings demonstrate an augmentation of hypoxic ventilatory reactivity at the carotid body level after peripheral pharmacological blockade of MAO, likely mediated by DA degradation through a switched-on COMT pathway. The inference would be that less endogenous DA augments ventilation, and *vice versa*, increased availability of DA dampens ventilation. Although the study design could not settle the exact role of DA in ventilatory chemoregulation, we believe the results may help understand the niceties of DA action. In addition, the findings of increased hypoxic ventilatory reactivity after the MAO inhibitor debrisoquine and no effect on ventilation of the COMT inhibitor entacapone, both clinically used drugs, may be of therapeutic consequence.

Conflicts of Interest The authors declare no conflicts of interest regarding this publication.

References

Bainbridge CW, Heistad D (1980) Effect of haloperidol on ventilatory responses to dopamine in man. J Pharmacol Exp Ther 213:13–17

Bee D, Pallot DJ (1995) Acute hypoxic ventilation, carotid body cell division, and dopamine content during early hypoxia in rats. J Appl Physiol 79:1504–1511

Bisgard GE, Mitchell RA, Herbert DA (1979) Effects of dopamine, norepinephrine and 5-hydroxytryptamine on the carotid body of the dog. Respir Physiol 37:61–80

Black AMS, Comroe JH, Jacobs L (1972) Species difference in carotid body response of cat and dog to dopamine and serotonin. Am J Physiol 223:1097–1102

Chow CM, Winder C, Read DJC (1986) Influence of endogenous dopamine on carotid body discharge and ventilation. J Appl Physiol 60:370–375

Eyzaguirre C, Zapata P (1984) Perspectives in carotid body research. J Appl Physiol 57:931–957

Faff L, Kowalewski C, Pokorski M (1999) Protein kinase C – a potential modifier of carotid body function. Monaldi Arch Chest Dis 54:172–177

Gonzalez C, Almaraz L, Obeso A, Rigual R (1994) Carotid body chemoreceptors: from natural stimuli to sensory discharges. Physiol Rev 74:829–898

Hanbauer I, Hellström S (1978) The regulation of dopamine and noradrenaline in the rat carotid body and its modification by denervation and by hypoxia. J Physiol Lond 282:21–31

Hellström S (1977) Putative neurotransmitters in the carotid body. Mass fragmentographic studies. Adv Biochem Psychopharmacol 16:257–263

Hovevey-Sion D, Kopin IJ, Stull RW, Goldstein DS (1989) Effects of monoamine oxidase inhibitors on levels of catechols and homovanillic acid in striatum and plasma. Neuropharmacology 28:791–797

Illi A, Sundberg S, Ojala-Karlsson P, Scheinin M, Gordin A (1996) Simultaneous inhibition of catechol-O-methyltransferase and monoamine oxidase A: effects on hemodynamics and catecholamine metabolism in healthy volunteers. Clin Pharmacol Ther 59:450–457

Joels N, White H (1968) The contribution of the arterial chemoreceptors to the stimulation of respiration by adrenaline in the cat. J Physiol Lond 197:1–23

Lahiri S, Nishino T, Mokashi A, Mulligan E (1980) Relative responses of aortic and carotid body chemoreceptors to hypotension. J Appl Physiol 48:781–788

Llados F, Zapata P (1978) Effects of dopamine analogues and antagonists on carotid body chemosensors in situ. J Physiol Lond 274:487–499

Lundberg DG, Breese G, Mueller R (1979) Dopaminergic interaction with the respiratory control system in the rat. Eur J Pharmacol 54:153–159

Matsumoto S, Nishimura Y, Kohno M, Nakajima T (1980) Effects of haloperidol on chemoreceptor reflex ventilatory response in the rabbit. Arch Int Pharmacodyn Ther 247:234–242

Monteiro TC, Obeso A, Gonzalez C, Monteiro EC (2009) Does ageing modify ventilatory responses to dopamine in anaesthetised rats breathing spontaneously? Adv Exp Med Biol 648:265–271

Monteiro TC, Batuca JR, Obeso A, Gonzalez K, Monteiro EC (2011) Carotid body function in aged rats: responses to hypoxia, ischemia, dopamine, and adenosine. Age 33:337–350

Smatresk NJ, Pokorski M, Lahiri S (1983) Opposing effects of dopamine receptor blockade on ventilation and carotid chemosensory activity. J Appl Physiol 54:1567–1573

Somers VK, Mark AL, Zavala DC, Abboud FM (1989) Contrasting effects of hypoxia and hypercapnia on ventilation and sympathetic activity in humans. J Appl Physiol 67:2101–2106

Tomares SM, Bamford OS, Sterni LM, Fitzgerald RS, Carroll JL (1994) Effects of domperidone on neonatal and adult carotid chemoreceptors in the cat. J Appl Physiol 77:1274–1280

Trendelenburg U (1984) The influence of inhibition of catechol-O-methyl transferase or of monoamine oxidase on the extraneuronal metabolism of 3H-(-)-noradrenaline in the rat heart. Naunyn Schmiedebergs Arch Pharmacol 327:285–292

Vicario I, Rigual R, Obeso A, Gonzalez C (2000) Characterization of the synthesis and release of catecholamine in the rat carotid body in vitro. Am J Physiol Cell Physiol 278:C490–C499

Wang Y, Berndt TJ, Gross JM, Peterson MA, So MJ, Knox FG (2001) Effect of inhibition of MAO and COMT on intrarenal dopamine and serotonin and on renal function. Am J Physiol Regul Integr Comp Physiol 280:R248–R254

Ward DS, Bellville JW (1982) Reduction of hypoxic ventilator drive by dopamine. Anesth Analg 61:333–337

Weyler W, Hsu YP, Breakefield XO (1990) Biochemistry and genetics of monoamine oxidase. Pharmacol Ther 47:391–417

Advs Exp. Medicine, Biology - Neuroscience and Respiration (2015) 6: 19–22
DOI 10.1007/5584_2014_70

Published online: 14 October 2014

Adaptation of Olfactory Threshold at High Altitude

R. Ruffini, C. Di Giulio, V. Verratti, M. Pokorski, G. Fanò-Illic, and A. Mazzatenta

Abstract

The aim of this study was to investigate the effects of the extreme environment of high altitude hypoxia on olfactory threshold. The study was conducted before, during, and after a scientific expedition to Mera Peak (5,800 m). The n-butanol test was used for the assessment of the magnitude of the olfactory threshold. The finding was that the olfactory threshold dramatically increased at high altitude. We conclude that there is a physiological adaptation of olfaction due to altitude-hypoxia.

Keywords

Altitude • Extreme environment • Hypoxia • Olfactory threshold • Smell

1 Introduction

Smell is a key sensorial system to identify food, environment, and it also is involved in such baffling processes as a partner selection. Nevertheless, olfaction is a 'neglected' sense because it is rarely investigated in medical practice, where patients are often unaware of olfactory alterations (Haddad et al. 2008; Doty and Laing 2003). Olfactory examination may give important information, since olfaction impairment may be associated with focal or diffuse disorders of the nervous system (Doty 2012; Graves et al. 1999). Moreover, olfactory ability changes with age; it begins to decline steadily in the 40s, frequently resulting in complete anosmia in elderly people (Doty et al. 1984a). In older people, impairment of olfactory function may represent an early diagnostic feature of a neurodegenerative disease (Mesholam et al. 1998).

Olfaction is a complex system (Haddad et al. 2008) composed of several physiological parameters, e.g. threshold, identification, and discrimination. In particular, the olfactory threshold is a measure of the lowest concentration of an odorant which activates the olfactory sensory

R. Ruffini, C. Di Giulio, V. Verratti, and G. Fanò-Illic
Department of Neuroscience & Imaging, University of Chieti-Pescara, Chieti, Italy

M. Pokorski
Department of Nursing, Public Higher Medical Professional School in Opole, 68 Katowicka St., 45-060 Opole, Poland
e-mail: m_pokorski@hotmail.com

A. Mazzatenta (✉)
Department of Neuroscience & Imaging, University of Chieti-Pescara, Chieti, Italy

University of Teramo and Neuroscience Institute CNR, Pisa, Italy

Istituto di Neuroscienze (IN), Consiglio Nazionale delle Ricerche (CNR), Via Moruzzi 1, Pisa 56100, Italy
e-mail: amazzatenta@yahoo.com

neurons and is one of the most investigated features (Cain 1982). Whereas the threshold testing is assumed to reflect mainly peripheral olfactory processes, discrimination and identification tasks seem to be more related to higher order cognitive processing, largely involving various memory functions (Landis et al. 2005). Olfactory function can be investigated using psychophysiological (Ehrenstein and Ehrenstein 1999), electrophysiological (Hummel and Kobal 2001), or imaging techniques (Cerf-Ducastel and Murphy 2003). In clinical practice, the University of Pennsylvania Smell Identification test (Doty et al. 1984b), the 'Sniffin' Sticks' test (Hummel et al. 1997), and the Connecticut Chemosensory Clinical Research Center test (CCCRC) (Cain et al. 1988) are the most commonly used psychophysiological tests for a comprehensive assessment of olfactory function. There are sparse reports on the effects of experimental hypobaric hypoxia on olfaction (Kühn et al. 2008, 2009; Semeria 1957), but olfaction has never been tested at true high altitude. Because of that and the known similarity between hypoxia and the aging process (Mazzatenta et al. 2013a), in the present study we set out to examine the possible influence of the extreme environment of high altitude hypoxia on olfactory threshold.

2 Methods

Seventeen healthy individuals of the mean age 45.3 ± 11.3 (SD) years (F/M – 5/12) were enrolled into the study during a scientific expedition to Mera Peak (5,800 m), taking 21 days at high altitude. The exclusion criteria were smoking, alcohol or drug consumption, impaired sense of smell, any overt pathology or disease, recent clinical surgery or anesthesia, and use of drugs. An additional exclusion criterion for women was the time between the 17 and 21st day of the menstrual cycle, to prevent the possible hormonal interference with the results. All experimental procedures were clearly explained, and the participants provided written informed consent prior to the testing sessions. The participants were free to interrupt the testing at any time in case of any difficult to manage or unpropitious feelings. The study was performed in accord with the ethical standards of the Helsinki Declaration of 2008 and was approved by a local Human Research Review Board and Ethical Committee. The olfactory threshold was evaluated by applying a standard n-butanol test (woody-alcohol odor), using a growing series of 9 M concentrations ranging from 10^{-5} to 0.6 M in the following ascending order: $9.14e^{-5}$, $2.74e^{-4}$, $8.23e^{-4}$, 0.00245, 0.0074, 0.022, 0.67, 0.2, and 0.6 M (Cain 1982). The trial ended when the subject perceived an olfactory sensation. The test was re-administered three times. The threshold value was defined as the mean score of dilution steps. The olfactory stimulation was performed at different altitudes: pre-test at sea level, test at 3,600 m, test at 5,800 m, and post-test at sea level. The solution was sniffed from a distance of 2 cm from the nose, for 3–4 s, and with an interval of at least 5 min between trials of different concentrations. The time interval between two stimuli of the same concentration was 30 s. The solution were presented in progressively higher concentrations until the volunteers perceived the odor, which corresponds to an individual olfactory threshold.

Data are presented as mean values ± SD. Statistical comparisons were made with one-way ANOVA. The alpha level was set at 0.05. Analysis was carried out using an SPSS ver. 17.0 commercial package.

3 Results

The mean pre-test olfactory threshold of the subjects was 5.71 ± 1.10 dilution steps (>0.05 % of n-butanol) before the expedition at the sea level (0 m). The threshold increased slightly to 5.82 ± 0.64 dilution steps at 3,600 m. The increase was distinctly greater at 5,800 m amounting to 6.94 ± 0.24 dilution steps (<0.43 % of n-butanol). The post-test result at the return to sea level was 5.06 ± 0.43 dilution steps, which was way below the beginning sea

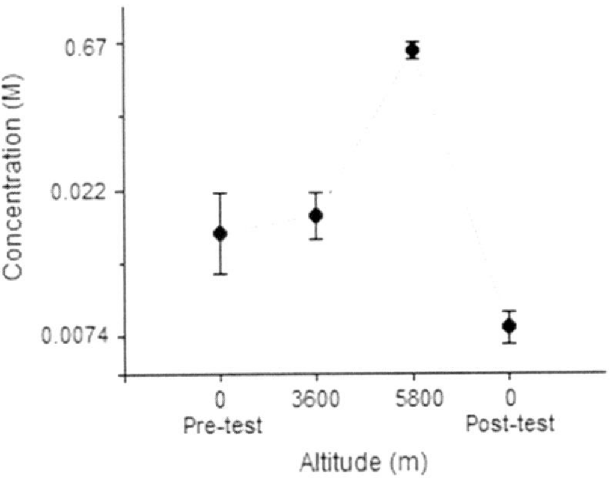

Fig. 1 The olfactory threshold, defined as the mean score of dilution steps: before the expedition at sea level (pre-test), altitudes of 3,600 and 5,800 m; and back at sea level (post-test). In the extreme environment of the higher altitude, the threshold increased by 1.5 concentration points. The olfactory threshold level showed undershoot in the post-test recovery compared with the pre-test baseline level

level (Fig. 1). One-way ANOVA showed a significant effect of altitude on the individual olfactory threshold $F_{(3,64)} = 22.3$, $p < 0.001$.

Differences in the distribution of the olfactory threshold frequencies per molar concentration of the stimulating solution at different altitudes are shown in Fig. 2. There was a very substantial reduction in frequency classes from the pre-test (0 m) to the high altitude of 3,600–5,800 m levels; signs of recovery were noted at the post-test sea level performed after 21 days' stay in the extreme environment.

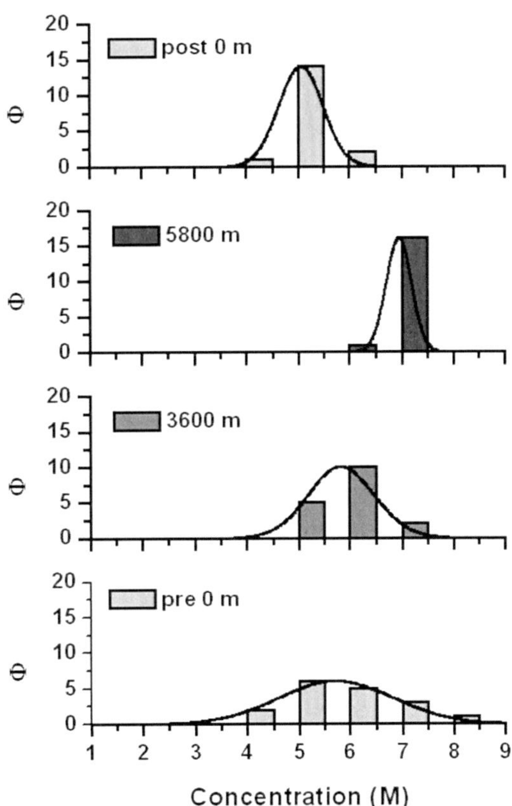

Fig. 2 Distribution of olfactory threshold frequencies per molar concentration of the stimulating solution at different altitudes: pre-test (0 m); 3,600 m; 5,800 m, and post-test (0 m)

4 Discussion

The major finding of the present study was that the perception of smell was distorted at high altitude. The distortion consisted of a dramatic increase in the olfactory threshold during the stay at the altitude of 5,800 m. Natural hypoxia is an interesting model to study the adaptation of olfactory threshold to extreme conditions. The increase in the olfactory threshold we observed from the pre-test to 3,600 m and to 5,800 m could be considered as a normal physical decrease of stimulus volatility linked to the altitude temperature changes. However, the post-test drop in threshold below the baseline pre-test sea level cannot be explained by such a simple justification. The post-test olfactory threshold clearly suggests a physiological mechanism that affects the periphery of the olfactory perception pathway. The adaptation of the olfactory threshold at high-altitude is, in all likelihood, produced by the accompanying hypoxia. The physiological mechanisms underlying this adaptation are as yet unknown. There is a literature report showing an interaction between the olfactory bulb and the reticular neurons of bulbar respiratory center in hypoxia (Karapetian et al. 2012). Interestingly, chronic exposure to hypoxia may underlie the appearance of some diseases, such as multiple chemical sensitivity syndrome that also affects olfaction

(Mazzatenta et al. 2013a, b). Altitude-hypoxia might be considered an important experimental model to investigate the effect of hypoxia on the olfactory physiology. It would be interesting to investigate the olfactory threshold also in the hypobaric chamber. Furthermore, such model could be useful in the understanding of the mechanisms of the aging process as well as neurodegenerative diseases.

Acknowledgements Thanks Josè Luiz Dantas for technical assistance.

Conflicts of Interest The authors declare no conflicts of interest in relation to this article.

References

Cain WS (1982) Sumner's 'on testing the sense of smell' revisited. Yale J Biol Med 55:515–519

Cain WS, Gent JP, Goodspeed RD, Leonard G (1988) Evaluation of olfactory dysfunction in the Connecticut Chemosensory Clinical Research Center. Laryngoscope 98:83–88

Cerf-Ducastel B, Murphy C (2003) FMRI brain activation in response to odors is reduced in primary olfactory areas of elderly subjects. Brain Res 986:39–53

Doty RL (2012) Olfactory dysfunction in Parkinson disease. Nat Rev Neurol 8:329–339

Doty RL, Laing DG (2003) Psychophysical measurements of human olfactory function, including odorant mixture assessment. In: Doty RL (ed) Handbook of olfaction and gustation, 2nd ed. Revised and expanded. Marcel Dekker, New York, pp 203–228

Doty RL, Shaman P, Applebaum SL, Giberson R, Sikorski L, Rosenberg L (1984a) Smell identification ability: changes with age. Science 226:1441–1443

Doty RL, Shaman P, Dunn M (1984b) Development of the University of Pennsylvania Smell Identification Test: a standardized microencapsulated test of olfactory function. Physiol Behav 32:489–502

Ehrenstein WH, Ehrenstein A (1999) Psychophysical methods. In: Windhorst U, Johansson H (eds) Modern techniques in neuroscience research, 1st edn. Springer, Berlin, pp 1211–1241

Graves AB, Bowen JD, Rajaram MS (1999) Impaired olfaction as a marker for cognitive decline. Interaction with apolipoprotein E ε4 status. Neurology 53:1480–1487

Haddad R, Lapid H, Harel D, Sobel N (2008) Measuring smells. Curr Opin Neurobiol 18:438–444

Hummel T, Kobal G (2001) Olfactory event-related potentials. In: Simon SA, Nicolelis MAL (eds) Methods and frontiers in chemosensory research, 1st edn. CRC Press, New York, pp 429–464

Hummel T, Sekinger B, Wolf SR, Pauli E, Kobal G (1997) Sniffin' sticks': olfactory performance assessed by the combined testing of odor identification, odor discrimination and odor threshold. Chem Senses 22:39–52

Karapetian MA, Adamian NI, Arutiunian RS (2012) The influence of bulbus olfactorius at the reticular neurons of bulbar respiratory center in hypoxia. Ross Fiziol Zh Im I M Sechenova 98:767–776

Kühn M, Welsch H, Zahnert T, Hummel T (2008) Changes of pressure and humidity affect olfactory function. Eur Arch Otorhinolaryngol 265:299–302

Kühn M, Welsch H, Zahnert T, Hummel T (2009) Is olfactory function impaired in moderate height? Laryngorhinootologie 88:583–586

Landis BN, Hummel T, Lacroix JS (2005) Basic and clinical aspects of olfaction. Adv Tech Stand Neurosurg 30:70–105

Mazzatenta A, Pokorski M, Cozzutto S, Barbieri P, Veratti V, Di Giulio C (2013a) Non-invasive assessment of exhaled breath pattern in patients with multiple chemical sensibility disorder. Adv Exp Med Biol 756:179–188

Mazzatenta A, Di Giulio C, Pokorski M (2013b) Pathologies currently identified by exhaled biomarkers. Respir Physiol Neurobiol 187:128–134

Mesholam RI, Moberg PJ, Mahr RN, Doty RL (1998) Olfaction in neurodegenerative disease: a meta-analysis of olfactory functioning in Alzheimer's and Parkinson's diseases. Arch Neurol 55:84–90

Semeria C (1957) Olfactory and gustatory changes in experimental hypoxemic conditions in man. Minerva Otorinolaringol 7:354–358

Advs Exp. Medicine, Biology - Neuroscience and Respiration (2015) 6: 23–33
DOI 10.1007/5584_2014_73

Published online: 14 October 2014

Guanosine Protects Glial Cells Against 6-Hydroxydopamine Toxicity

Patricia Giuliani, Patrizia Ballerini, Silvana Buccella, Renata Ciccarelli, Michel P. Rathbone, Silvia Romano, Iolanda D'Alimonte, Francesco Caciagli, Patrizia Di Iorio, and Mieczyslaw Pokorski

Abstract

Increasing body of evidence indicates that neuron-neuroglia interaction may play a key role in determining the progression of neurodegenerative diseases including Parkinson's disease (PD), a chronic pathological condition characterized by selective loss of dopaminergic (DA) neurons in the substantia nigra. We have previously reported that guanosine (GUO) antagonizes MPP^+-induced cytotoxicity in neuroblastoma cells and exerts neuroprotective effects against 6-hydroxydopamine (6-OHDA) and beta-amyloid-induced apoptosis of SH-SY5Y cells. In the present study we demonstrate that GUO protected C6 glioma cells, taken as a model system for astrocytes, from 6-OHDA-induced neurotoxicity. We show that GUO, either alone or in combination with 6-OHDA activated the cell survival pathways ERK and PI3K/Akt. The involvement of these signaling systems in the mechanism of the nucleoside action was strengthened by a reduction of the protective effect when glial cells were pretreated with U0126 or LY294002, the specific inhibitors of MEK1/2 and PI3K, respectively. Since the protective effect on glial cell death of GUO was not affected by pretreatment with a cocktail of nucleoside transporter blockers, GUO transport and its intracellular accumulation were not at play in our *in vitro* model of PD. This fits well with our data which pointed

P. Giuliani, S. Buccella, R. Ciccarelli, S. Romano, I. D'Alimonte, F. Caciagli, and P. Di Iorio
Department of Experimental and Clinical Sciences, "G. d'Annunzio" University, Chieti-Pescara, 29 via dei Vestini, NPD PalB, 66013 Chieti, Italy

P. Ballerini (✉)
Department of Psychological, Humanities and Territory Sciences, School of Medicine and Health Sciences, "G. d'Annunzio" University, Chieti-Pescara, 29 via dei Vestini, NPD PalB, 66013 Chieti, Italy
e-mail: p.ballerini@unich.it

M.P. Rathbone
Department of Medicine, McMaster University, Health Sciences Center, Hamilton, ON, Canada

M. Pokorski (✉)
Department of Nursing, Public Higher Medical Professional School in Opole, 68 Katowicka St., 45-060 Opole, Poland

Institute of Psychology, Opole University, 1 Plac Staszica, 45-052 Opole, Poland
e-mail: m_pokorski@hotmail.com

to the presence of specific binding sites for GUO on rat brain membranes. On the whole, the results described in the present study, along with our recent evidence showing that GUO when administered to rats *via* intraperitoneal injection is able to reach the brain and with previous data indicating that it stimulates the release of neurotrophic factors, suggest that GUO, a natural compound, by acting at the glial level could be a promising agent to be tested against neurodegeneration.

Keywords
Apoptosis • Dopamine • Gial cells • Guanosine • Neurodegeneration • Parkinson's disease • Substantia nigra

1 Introduction

Parkinson's disease (PD) is a chronic and progressive neurodegenerative pathological condition characterized by selective loss of dopaminergic (DA) neurons in the substantia nigra (Dauer and Przedborski 2003; Birkmayer and Hornykiewicz 1961). Although different mechanisms have been considered influential for the PD pathogenesis, including inflammation, oxidative stress, excitotoxicity, protein misfolding, or apoptosis, none has been indicated as the primary cause of the disease, since all of them may likely act in a complex integrated pathway to promote neurodegeneration (Fujita et al. 2013; Barnum and Tansey 2010). Therapies for PD are mainly based on DA replacement by L-3,4-dihydroxyphenylalanine (L-DOPA), an approach which is effective in reducing motor handicap and alleviating the disease-associated depression and pain. However, chronic administration of L-DOPA often causes motor and psychiatric side effects which are reported to be as debilitating as PD itself (Andrew et al. 1993; Curtius et al. 1974).

In vitro treatment of cells with 6-hydroxydopamine (6-OHDA) and *in vivo* administration of this neurotoxin to laboratory animals represent one of the most used experimental models of PD. Moreover, 6-OHDA content is increased in the brain and urine of patients with PD treated with L-DOPA (Andrew et al. 1993; Curtius et al. 1974) and a role for this neurotoxin in the pathogenesis of the disease has been suggested.

We have recently reported that guanosine (GUO) protected SH-SY5Y neuroblastoma cells when exposed to neurotoxins, including MPP^+ and 6-OHDA (Giuliani et al. 2012b; Pettifer et al. 2007). We have also reported that in a clinically relevant chronic animal model of PD, developed by using a proteasome inhibitor (McNaught et al. 2004), GUO treatment: (i) reduced apoptosis; (ii) increased tyrosine hydroxylase positive DA neurons; and (iii) ameliorated symptoms (Su et al. 2009). Increasing body of evidence indicates that neuron-neuroglia interaction may play a key role in determining the progression of neurodegenerative diseases, including PD (McGeer and McGeer 2008; Mena and Garcia de Yebenes 2008).

6-OHDA, beside its extracellular-mediated cytotoxic effects, damages catecholaminergic neurons following its uptake into the cells by DA transporters (Blum et al. 2000). Both monoamine oxidase (MAO) and catechol-O-methyltransferase (COMT) are present in astroglial cells (Fitzgerald et al. 1990; Hansson and Sellstrom 1983; Pelton et al. 1981), suggesting that astrocyte uptake systems are likely to play an important role. It has been reported that not only catecholaminergic neurons but also astrocytes are able to take up DA by high-affinity Na^+-dependent and Na^+-independent systems (Pelton et al. 1981) and by EMT, which is an extraneuronal Na^+-independent monoamine transporter system (Inazu et al. 1999a, b).

The role of astrocytes in DA neurons' function is of interest as the glia/neurons ratio in the substantia nigra is the lowest in the brain (Damier et al. 1996; Makar et al. 1994; Sagara et al. 1993). This suggests that DA neurons could less rely on glial cells in case of different kinds of insults and that agents able to damage glial cells may contribute to the onset and progression of disease. Therefore, given the emerging role of astrocytes in the pathophysiology of PD and the need of evaluating novel therapeutic strategies for this disorder, in the present study we used C6 glioma cell cultures to investigate the neuroprotective effects of GUO in 6-OHDA-induced neurotoxicity.

2 Methods

The study was approved by the Ethics Committee of Chieti University in Italy.

2.1 Agents

6-OHDA, cells media, L-glutamine, poly-L-lysine, GUO, 6-[(4-nitrobenzyl)thio]-9-β-D-ribofuranosylpurine (NBTI), propentofylline (PPF), dipyridamole (DYP), 2-(4-morpholinyl)-8-phenyl-1(4H)-benzopyran-4-one hydrochloride (LY294002), and 1,4-diamino-2,3-dicyano-1,4-bis(o-aminophenylmercapto)butadiene monoethanolate (U0126) were purchased from Sigma Aldrich (Milano, Italy). Fetal bovine serum (FBS), penicillin/streptomycin (10,000 units/ml penicillin G sodium and 10,000 μg/ml streptomycin sulfate in 0.85 % saline), trypsin 10X liquid were obtained from Invitrogen (Milano, Italy). Antibody against phosphorylated-Akt (Ser473) and ERK1/2 were purchased from Cell Signaling Technology (CELBIO S.p.A, Milano, Italy). Anti-Actin (I-19) antibody was obtained from Santa Cruz Biotechnology (D.B.A. Italia S.r.l., Milano, Italy). Donkey anti-rabbit HPR-conjugated and chemiluminescence (ECL) detection kit was purchased from GE Healthcare (Milano, Italy).

2.2 C6 Glioma Cell Cultures

C6 glioma cells, were grown in DMEM containing 10 % FBS and 1 % penicillin/streptomycin in a humidified 5 % CO_2 incubator at 37 °C. The culture medium was changed every other day and cells prepared at an appropriate density depending on the kind of experiment to be performed. In all the experiments, GUO was dissolved in NaOH 0.1 M and added to the culture medium at a final concentration of 0.001 M NaOH. Some cells were exposed to 0.001 M NaOH, but, as previously reported by Pettifer et al. (2007), no modification was observed.

2.3 Cell Viability Assay

Cell viability was evaluated by the MTS assay using the CellTiter 96 AQueous One Solution Cell Proliferation Assay kit (Promega, Milano, Italy). The assay is based on the cleavage of the yellow tetrazolium salt [3-(4,5-dimethylthiazol-2-yl)-5-(3-carboxymethoxyphenyl)-2-(4-sulfophenyl)-2H-tetrazolium] to purple formazan crystals by metabolically active cells. Cultured cells were starved by serum removal for 24 h and then treated with 6-OHDA at different concentrations (ranging from 5 to 200 μM) for 24 h. The EC_{50} value was used to treat the cells in combination with 300 μM GUO (co-treatment).

Inhibitors of the nucleoside transporters (10 μM NBTI plus 100 μM PPF plus 10 μM DYP) were added to the medium 1 h before the 6-OHDA/GUO co-treatment until the end of the experiment. After treatment, 20 μl of the One-Solution Reagent were added to each well and the plates were incubated at 37 °C for 2 h. The optical density at 490 nm was measured using a Packard SpectraCount™ microplate reader. The value of viability of treated cells was expressed as a percentage of that from the corresponding control cells.

2.4 Hoechst 33258 Staining

After treatment with 30 μM 6-OHDA in the presence or absence of 300 μM GUO for 24 h, the cells were harvested and fixed for 30 min with gentle agitation in pre-chilled phosphate buffered saline (PBS) containing 4 % paraformaldehyde. After fixation at room temperature, C6 glioma cells were washed with pre-chilled PBS and then exposed to Hoechst 33258 2 mg/l in PBS at room temperature for 5 min. After washing, all samples were analyzed with a fluorescence microscope (Leica DM RXA2).

2.5 DNA Fragmentation

Specific apoptotic DNA fragmentation was evaluated by measuring the amount of cytosolic oligonucleosomes using a Cell Death ELISA kit (Roche Molecular Biochemicals; Milano, Italy) and it was carried out according to the manufacturer's instruction. After the exposure of the culture to 30 μM 6-OHDA alone or in combination with 300 μM GUO, cells were isolated for analysis of DNA fragmentation as described by Pettifer et al. (2007). Briefly, 10,000 viable cells/treatment were lysed and centrifuged to isolate fragmented oligonucleosomal DNA. The cytosolic fractions of cell lysates were transferred into streptavidin-coated microplate wells, and a mixture of biotin-linked anti-histone antibody and peroxidase-linked anti-DNA antibody was added and incubated for 2 h at room temperature. Plates were washed with the incubation buffer to remove the unfixed anti-DNA antibody and the peroxidase activity was determined spectrophotometrically with 2,2′-azino-bis[3-ethylbenzthiazoline-6-sulfonic acid] (ABTS) as substrate (absorbance of 405 nm). The amount of DNA fragmentation was expressed as a percentage of the positive control provided with the kit.

2.6 Western Blot Analysis

Western blot analysis was used to detect phosphorylated Akt (p-Akt) and ERK1/2 (p-ERK1/2). Cultured cells were rendered quiescent by serum removal for 24 h and treated with 30 μM 6-OHDA and 300 μM GUO alone or in combination. Inhibitors of nucleoside transporters (10 μM NBTI plus 100 μM PPF plus 10 μM DYP) were added to the medium 1 h before the 6-OHDA/GUO co-treatment until the end of experiment. At the end of the treatment, cells were washed twice with ice-cold PBS and then harvested at 4 °C in a lysis buffer (25 mM Tris buffer, pH 7.4, containing 150 mM NaCl, 100 μM sodium orthovanadate, 1.5 mM $MgCl_2$, 1.0 mM EDTA, 1.0 mM EGTA, 1 % NP40, 10 % glycerol, 1 mM PMSF, 5 μg/ml leupeptin, 5 μg/ml aprotinin). After 20 min on ice, cells were centrifuged at 14,000 rpm for 10 min at 4 °C. Supernatant aliquots were used for the determination of protein concentrations by the method of Bradford (Bio-Rad; Milano-Italy).

Proteins were diluted in Laemli-SDS sample buffer and boiled for 5 min. Equal amounts of proteins were loaded onto each lane of SDS-PAGE gel (12 % resolving gel, 4 % stacking gel) and resolved at 120 V constant. Gels were transferred onto PVDF membrane (Bio-Rad, Milano-Italy) at 100 V constant for 90 min at room temperature, and membranes were blocked in blocking buffer (PBS, 0.1 % Tween-20 with 5 % w/v non-fat dry milk) for 2 h. Blots were incubated overnight at 4 °C with specific primary antibodies. All primary antibodies were diluted 1:1,000 in primary antibody dilution buffer (PBS, 0.1 % Tween-20 with 2,5 % w/v non-fat dry milk) except for β-actin that was diluted 1:1,500. After washing three times for 15 min with wash buffer (PBS, 0.1 % Tween-20), membranes were exposed to a secondary antibody diluted 1:2,500 for 1 h at room temperature. The immunocomplexes were visualized using an enhancing ECL detection system. Densitometric analysis was performed for the quantification of the immunoblots using the Quantity One 1-D Analysis software (Bio-Rad; Milano, Italy).

2.7 Statistical Analysis

Data were analyzed with a two-tailed *t*-test and expressed as means ± SE. All experiments were performed at least three times. $P < 0.05$ was considered statistically significant.

3 Results

3.1 Effect of 6-OHDA on C6 Glioma Cell Viability

C6 glioma cells were treated with various concentrations of 6-OHDA (5–200 μM) for 24 h and cell viability was evaluated by conventional MTS reduction assay. As shown in Fig. 1, the proportion of viable cells was reduced by 6-OHDA in a concentration-dependent manner. The half maximal effective concentration (EC_{50}) of approximately 30 μM 6-OHDA was used in the experiments aimed at studying the protective effects of GUO.

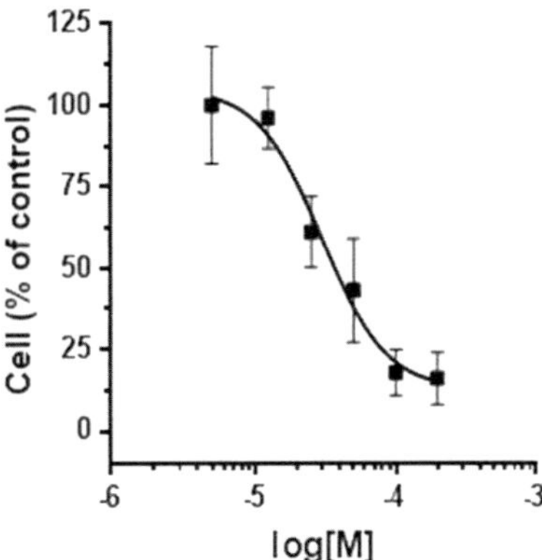

Fig. 1 Concentration response curve of 6-OHDA on C6 glioma cell viability. The curve was generated to identify the EC_{50} value for the toxin on cell survival. Data are means ± SE of at least six independent experiments and are expressed as a percentage of the untreated control group

3.2 Protective Effect of Guanosine on 6-OHDA-Mediated Toxicity

C6 glioma cells were treated with 30 μM 6-OHDA for 24 h in the presence or absence of 300 μM GUO. The concentration of the guanine nucleoside was chosen on the basis of our previous results showing its neuroprotective effects in different cell types, including glial and neuroblastoma cells (Di Iorio et al. 2004; Pettifer et al. 2004). The addition of 30 μM 6-OHDA led to the expected reduction in C6 glioma cell viability (Fig. 2). GUO (300 μM) added along with the toxin (co-treatment) effectively attenuated (by about 60 %) the 6-OHDA-induced cytotoxicity as determined by MTS reduction assay (Fig. 2).

To evaluate whether the protective effects of GUO were mediated by intracellular mechanisms following the nucleoside uptake, the cells were pretreated for 1 h with a cocktail of known nucleoside transporter blockers (10 μM NBTI, 100 μM PPF, and 10 μM DYP) (Parkinson et al. 2006). As shown in Fig. 2, the protective effect of GUO was not affected by cell pretreatment with these drugs.

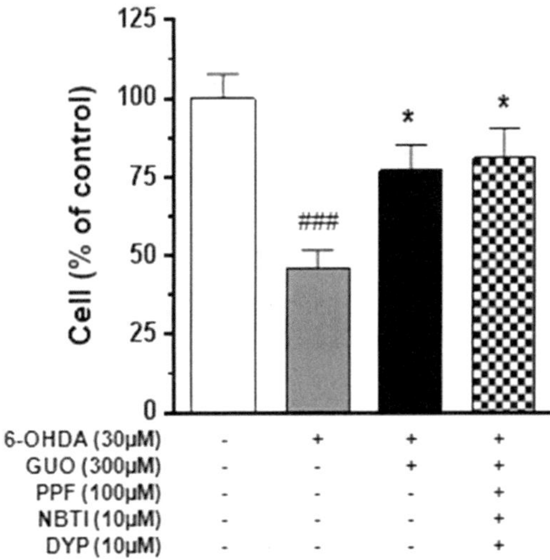

Fig. 2 Effect of guanosine (GUO) on C6 glioma cell cytotoxicity induced by 6-hydroxydopamine (6-OHDA). The effect of 300 μM guanosine on the toxicity induced by 30 μM 6-OHDA for 24 h was evaluated in the presence or absence of nucleoside transporter blockers. The blocker cocktail [10 μM 6-[(4-nitrobenzyl) thio]-9-β-D-ribofuranosylpurine (NBTI), 100 μM propentofylline (PPF), and 10 μM dipyridamole (DYP)] was added to the culture medium 1 h before 6-OHDA/GUO co-treatment. The C6 cell viability was determined by MTS assay as described in Methods. Each column represents the mean ± SE of at least six independent experiments. ###$P < 0.001$ *vs*. untreated cells (control); *$P < 0.05$ *vs*. 6-OHDA-treated cells

3.3 6-OHDA -Induced Apoptotic Death in C6 Glioma Cells

To determine the capability of the chosen 6-OHDA concentration to cause apoptosis,

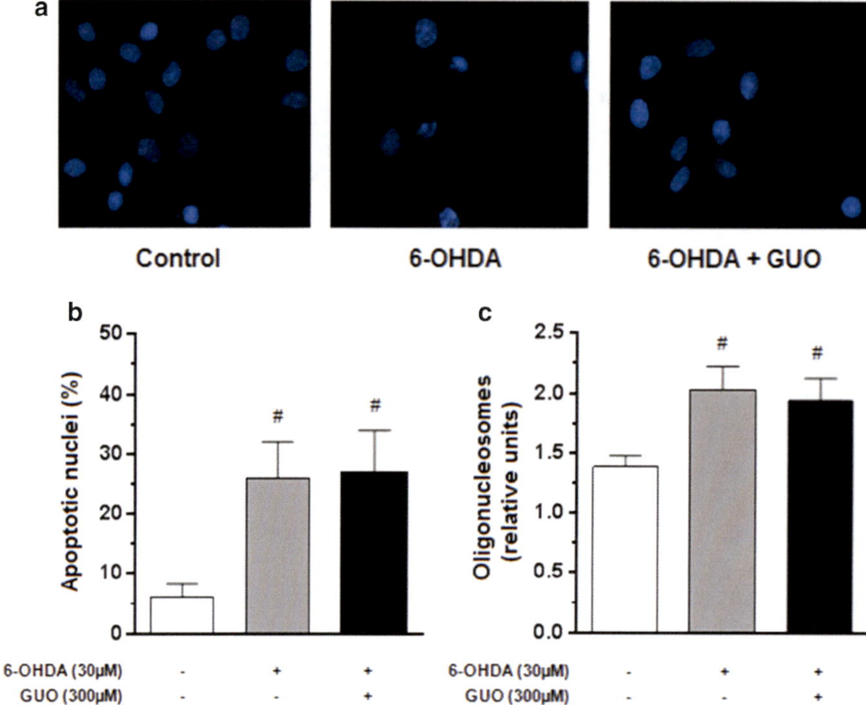

Fig. 3 Effect of 6-hydroxydopamine (6-OHDA) alone or in combination with guanosine (GUO) on C6 glioma cell apoptosis. Cultured cells were exposed to 30 μM 6-OHDA for 24 h in the presence or absence of 300 μM GUO. Untreated (control) and treated cells were stained with the fluorescent nuclear dye Hoechst 33258 (*Panel A*) and the statistical analysis of apoptotic cells was reported (*Panel B*). The apoptosis was also evaluated measuring the oligonucleosome formation by ELISA assay (*Panel C*). Photomicrographs are representative results taken from ten different fields from randomly selected slides. Each column value represents the mean ± SE of six independent experiments. $^{\#}p < 0.05$ *vs.* untreated cells (control)

DNA-sensitive dye Hoechst 33258 staining and DNA fragmentation assay by oligonucleosomal ELISA were used. The Hoechst 33258 staining was used to assess changes in nuclear morphology following cell treatment with 30 μM 6-OHDA for 24 h in the presence or absence of 300 μM GUO. As shown in Fig. 3A, nuclei in normal C6 glioma cells exhibited diffused Hoechst 33258 staining of chromatin. In contrast, about 27 % of nuclei in cells treated with 6-OHDA showed condensed chromatin (Fig. 3A, B). No protection was reported when 300 μM GUO was added to the cell medium along with the neurotoxin.

These results were confirmed by DNA fragmentation assay. 6-OHDA caused a 46 % increase in oligonucleosmal formation (Fig. 3C) and GUO (300 μM) added concomitantly with 6-OHDA was unable to reduce the toxin-mediated oligonucleosome formation (Fig. 3C).

3.4 ERK and PI3K/Akt Pathways Involvement in Neuroprotection by Guanosine Against 6-OHDA-Mediated Toxicity in C6 Glioma Cells

It has been reported that GUO induces a rapid increase in ERK1/2 and Akt phosphorylation in different cell types including PC12, microglia, and astrocytes (D'Alimonte et al. 2007; Di Iorio et al. 2004; Pettifer et al. 2004). To evaluate the involvement of the above mentioned cell survival pathways in GUO-mediated neuroprotection, we examined the effect of LY294002

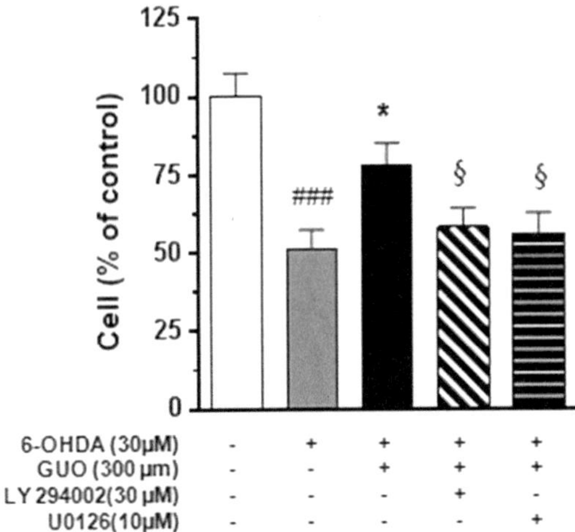

Fig. 4 Effects of PI3-kinase inhibitor (LY294002) and MEK1/2 inhibitor (U0126) on the protective effect of 300 μM guanosine (GUO) on 6-hydroxydopamine (6-OHDA)-induced toxicity. LY294002 (30 μM) or U0126 (10 μM) were added 1 h prior to the co-treatment with 300 μM GUO plus 30 μM 6-OHDA for 24 h. The C6 cell viability was determined by MTS assay as described in Methods. Each column represents the mean ± SE of at least six independent experiments. ###$P < 0.001$ *vs.* untreated cells (control); *$P < 0.05$ *vs.* 6-OHDA-treated cells; §$P < 0.05$ *vs.* the 6-OHDA/GUO co-treatment

(30 μM) an inhibitor of phosphoinositide-3-kinase (PI3K), an upstream of Akt, and U0126 (10 μM), an inhibitor of mitogen-activated protein kinase (MEK1/2), in glioma cell cultures co-treated with 30 μM 6-OHDA for 24 h. As shown in Fig. 4, both LY294002 and U0126 significantly reduced the effect of GUO on 6-OHDA-induced cytotoxicity as determined by MTS reduction assay. Consistent with the results obtained by MTS reduction assay, Western blotting analysis showed that 300 μM GUO was able to increase p-ERK1/2 and p-Akt – immunoreactivity in C6 glioma cells treated with the nucleoside for 15 min (Fig. 5A, C). As already reported for other cell types, including C6 glioma cells (Lee et al. 2011), also 6-OHDA increased p-ERK immunoreactivity (Fig. 5A).

It has been reported that sustained activation of ERK1/2 does not protect against either L-DOPA- or 6-OHDA-induced cytotoxicity (Jin et al. 2010). Thus, in an attempt to better define the characteristics of ERK activation mediated by co-treatment of C6 glioma cells with 30 μM 6-OHDA plus 300 μM GUO, analysis of the kinetics of ERK phosphorylation was performed. As shown in Fig. 5B, co-treatment significantly increased p-ERK1/2 level within 5 min, peaking at 15 min after administration, and strongly decreased it by 60 min.

Contrary to what was found for ERK1/2 phosphorylation, the neurotoxin was unable to enhance the levels of p-Akt in the same cells (Fig. 5C), whereas following the co-treatment with GUO and 6-OHDA, p-Akt levels increased by 39 % compared with the control values. The effect of cell co-treatment with 300 μM GUO and 30 μM 6-OHDA on p-ERK1/2 and p-Akt immunoreactivity was not affected by C6 glioma cell pretreatment with the cocktail of the nucleoside transporter blockers reported above (10 μM NBTI, 100 μM PPF, and 10 μM DYP) (Fig. 6A, B).

4 Discussion

In the present study we demonstrate that 6-OHDA, a neurotoxin used to induce experimental models of PD, caused cytotoxicity in a concentration-dependent manner in C6 glioma cells, taken as a model system for astrocytes. We also report that 300 μM GUO concomitantly administered with 6-OHDA counteracted the toxin-induced loss of viability in this *in vitro* model of PD. It has been shown that GUO and its metabolic product guanine are taken up into both neurons and astrocytes mainly *via* the equilibrative nucleoside transporter (Parkinson et al. 2006). Thus, to determine whether GUO-mediated cytoprotection was due to intracellular effects following its uptake by the cells, a cocktail of uptake inhibitors was used. In agreement with previous findings observed in primary cultures of astrocytes (Di Iorio et al. 2004), the results show that the transport of GUO and its consequent intracellular accumulation was not required for protecting the cells in our *in vitro* model of PD. This fits well with our previous data which pointed to the presence of specific binding sites for GUO on rat brain membranes (Traversa et al. 2002, 2003).

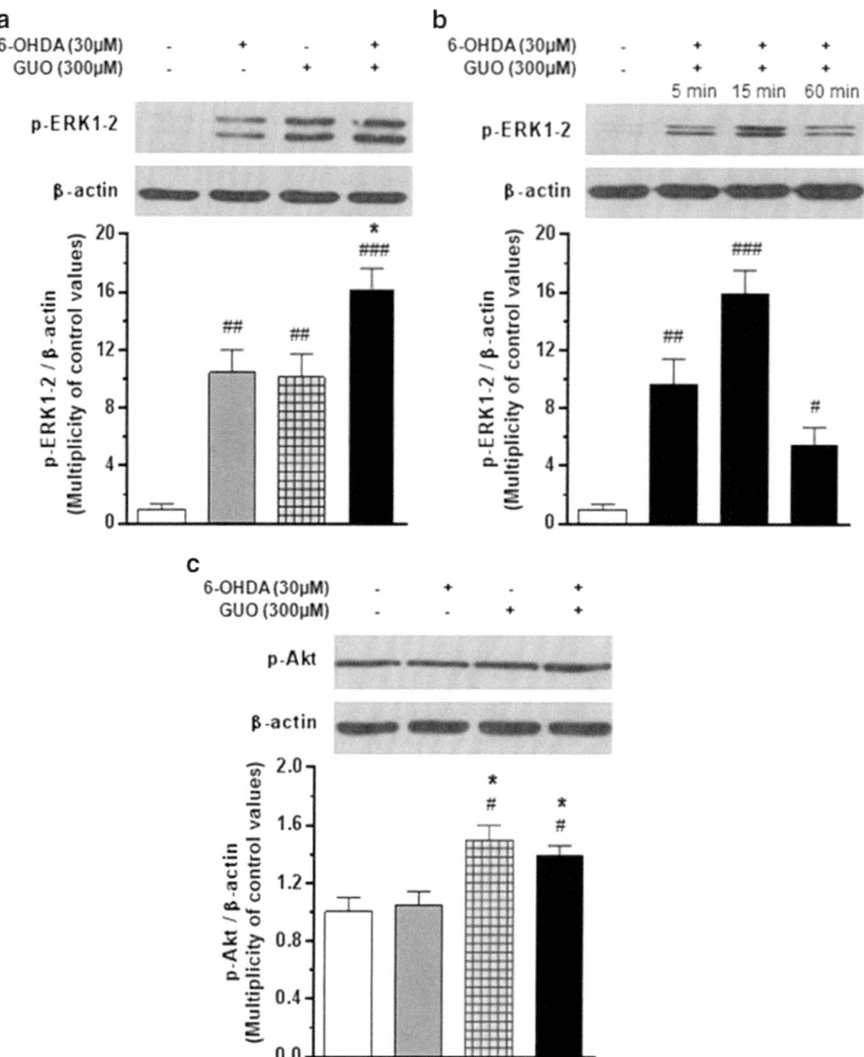

Fig. 5 Effect of 6-hydroxydopamine (6-OHDA), guanosine (GUO), or the combined treatment on the phosphorylation of ERK1/2 (*Panels A & B*) and Akt (*Panel C*) in C6 glioma cells. Cultured cells were incubated for 15 min with 30 μM 6-OHDA, 300 μM (GUO) or 6-OHDA/GUO co-treatment and harvested for western blot analysis. The protein expression of phosphorylated ERK1/2 (p-ERK1/2) or Akt (p-Akt), and β-actin were determined by using specific antibodies as described in Methods. To evaluate the time course of ERK1/2 activation induced by the co-treatment, cells were incubated for 5, 15, and 60 min with 30 μM 6-OHDA plus 300 μM GUO (*Panel B*). For each signaling pathway, a representative immunoblot is shown. Each column represents the mean ± SE of three independent experiments. #$P < 0.05$, ##$P < 0.01$, ###$P < 0.001$ *vs.* untreated cells (control); *$P < 0.05$ *vs.* 6-OHDA treated cells

The mechanisms underlying neurodegenerative processes in PD are still unclear. However, increasing evidence indicates that apoptosis is involved in the loss of dopaminergic neurons. We recently reported that GUO (300 μM) protected SH-SY5Y neuroblastoma cells when they were exposed to 6-OHDA, promoting their survival by (i) reducing the neurotoxin-mediated activation of p-38 and JNK; (ii) causing an early increase in phosphorylation of the anti-apoptotic kinase Akt; (iii) inactivating the pro-apoptotic factor GSK3β; and (iv) increasing the expression of the anti-apoptotic Bcl-2 protein (Giuliani et al. 2012b).

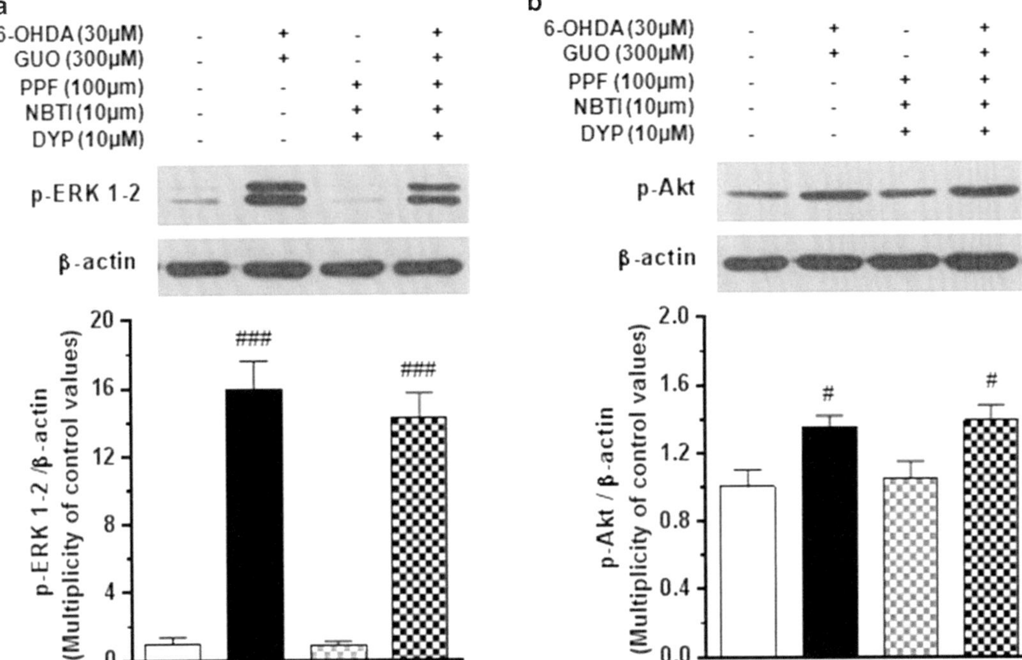

Fig. 6 Effect of the nucleoside transporter blockers on the phosphorylation of ERK1/2 (*Panel A*) and Akt (*Panel B*) induced by the C6 cell co-treatment with 6-hydroxydopamine (6-OHDA)/guanosine (GUO). Cultured cells were incubated for 15 min with 30 μM 6-OHDA plus 300 μM GUO. When used, the blocker cocktail [10 μM 6-[(4-nitrobenzyl)thio]-9-β-D-ribofuranosylpurine (NBTI), 100 μM propentofylline (PPF), and 10 μM dypiridamole (DYP)] was added to the culture medium 1 h before 6-OHDA/GUO cotreatment. The protein expression of phosphorylated ERK1/2 (p-ERK1/2) or Akt (p-Akt) and β-actin were determined by using specific antibodies as described in Methods. For each signaling pathway, a representative immunoblot is shown. Each column represents the mean ± SE of three independent experiments. #$P < 0.05$, ###$P < 0.001$ *vs.* untreated cells (control)

Apoptotic cells can be recognized by characteristic morphological changes, which are similar across cell types and species (Häcker 2000; Leist et al. 1997). The assessment of cell death by using Hoechst 33252 staining and oligonulceosomal formation is a combination of simple and reproducible methods to distinguish among viable and apoptotic cells. In the present work, we found that 6-OHDA induced apoptosis in C6 glioma cell but, unlike to what has been reported in neuroblastoma cells, GUO was unable to protect glial cultures against this kind of toxin-induced cell death.

We also reported that 6-OHDA induced a rapid phosphorylation of ERK1/2. Activation of the ERK pathway is known to contribute to neuronal cell survival in different models of neurotoxicity (Jin et al. 2002; Abe and Saito 2000; Han and Holtzman 2000), whereas sustained ERK phosphorylation has been shown to be involved in apoptosis processes and in a decrease of DA biosynthesis (Kulich and Chu 2001). Thus, our results are consistent with an attempt of glial cells to exert a defense response against the toxin. As our data indicate that GUO *per se* is able to cause a rapid and transient ERK phosphorylation, we hypothesize that the regulation of ERK phosphorylation might be involved in the mechanisms of GUO-induced cell survival.

Neuronal lesions caused by 6-OHDA are complex and the toxin can induce apoptosis and pre-necrotic lesions (Mayo et al. 1999). GUO could prevent the incipient signs of necrosis by enabling the cells to activate the recovery mechanisms. When the nucleoside was added to cultures along with 6-OHDA, a more intense early ERK phosphorylation, compared with that induced by the toxin alone, was observed. This

ERK activation decreased toward the baseline level at 60 min. Inhibition of this pathway by U0126 significantly attenuated the protective effect of GUO as shown by the MTS reduction assay. This finding strengthens the notion that the ERK pathway might play a role in inducing survival of astrocytes by GUO.

Another finding of the present study was that 6-OHDA did not evoke phosphorylation of Akt, a key mediator downstream of PI3 kinase reported to be involved in several physiological cellular processes including metabolic cell responses and cell survival, GUO, on the other hand, increased p-Akt levels either when administered alone or in combination with the neurotoxin. When the PI3K/Akt pathway was blocked with the specific pharmacological inhibitor LY294002, we found a marked reduction of the protective action of GUO evaluated by MTS reduction assay, suggesting that also this protein kinase is involved in the mechanism necessary for the nucleoside to protect glia against 6-OHDA-induced cell death. Alterations in PI3k/Akt signaling have been associated with loss of DA neurons in PD (Timmons et al. 2009) and recently the ability of Myr-Akt to induce a robust, accurate, and functionally integrated axon regrowth in nigral DA neurons in a 6-OHDA animal model of PD has been demonstrated. These observations suggest that targeting the PI3K-Akt pathway may represent a pharmacological approach of a therapeutic interest in neurodegenerative disorders including PD.

The effects of GUO on both ERK and Akt phosphorylation, when administered in combination with 6-OHDA, were not affected by C6 glioma cell pretreatment with a cocktail of nucleoside uptake inhibitors, lending support for a notion that GUO acts through the activation of membrane receptor sites. A work is in progress in our laboratories to deorphanize a G-coupled orphan receptor which seems a good candidate as a novel purinergic receptor for guanine-based purines. We have also recently reported that GUO administered intraperitoneally to rats is able to pass the blood-brain barrier (Giuliani et al. 2012a) and stimulates the release of neurotrophic factors (Di Iorio et al. 2004). Taken together the present findings demonstrate that GUO might be of research interest as a novel therapeutic approach for treatment of PD.

Acknowledgments PG and PB contributed to this work equally.

Conflicts of Interest The authors declare no conflicts of interest in realtion to this article.

References

Abe K, Saito H (2000) Neurotrophic effect of basic fibroblast growth factor is mediated by the p42/p44 mitogen-activated protein kinase cascade in cultured rat cortical neurons. Brain Res Dev Brain Res 122:81–85

Andrew R, Watson DG, Best SA, Midgley JM, Wenlong H, Petty RK (1993) The determination of hydroxydopamines and other trace amines in the urine of Parkinsonian patients and normal controls. Neurochem Res 18:1175–1177

Barnum CJ, Tansey MG (2010) Modeling neuroinflammatory pathogenesis of Parkinson's disease. Prog Brain Res 184:113–132

Birkmayer W, Hornykiewicz O (1961) The L-3,4-dihydroxyphenylalanine (DOPA)-effect in Parkinson-akinesia. Wien Klin Wochenschr 73:787–788

Blum D, Torch S, Nissou MF, Benabid AL, Verna JM (2000) Extracellular toxicity of 6-hydroxydopamine on PC12 cells. Neurosci Lett 283:193–196

Curtius HC, Wolfensberger M, Steinmann B, Redweik U, Siegfried J (1974) Mass fragmentography of dopamine and 6-hydroxydopamine. Application to the determination of dopamine in human brain biopsies from the caudate nucleus. J Chromatogr 99:529–540

D'Alimonte I, Ciccarelli R, Di Iorio P, Nargi E, Buccella S, Giuliani P, Rathbone MP, Jiang S, Caciagli F, Ballerini P (2007) Activation of P2X(7) receptors stimulates the expression of P2Y(2) receptor mRNA in astrocytes cultured from rat brain. Int J Immunopathol Pharmacol 20:301–316

Damier P, Kastner A, Agid Y, Hirsch EC (1996) Does monoamine oxidase type B play a role in dopaminergic nerve cell death in Parkinson's disease? Neurology 46:1262–1269

Dauer W, Przedborski S (2003) Parkinson's disease: mechanisms and models. Neuron 39:889–909

Di Iorio P, Ballerini P, Traversa U, Nicoletti F, D'Alimonte I, Kleywegt S, Werstiuk ES, Rathbone MP, Caciagli F, Ciccarelli R (2004) The antiapoptotic effect of guanosine is mediated by the activation of the PI 3-kinase/AKT/PKB pathway in cultured rat astrocytes. Glia 46:356–368

Fitzgerald LW, Kaplinsky L, Kimelberg HK (1990) Serotonin metabolism by monoamine oxidase in rat primary astrocyte cultures. J Neurochem 55:2008–2014

Fujita KA, Ostaszewski M, Matsuoka Y, Ghosh S, Glaab E, Trefois C, Crespo I, Perumal TM, Jurkowski W, Antony PM, Diederich N, Buttini M, Kodama A, Satagopam VP, Eifes S, Del Sol A, Schneider R, Kitano H, Balling R (2013) Integrating pathways of Parkinson's disease in a molecular interaction map. Mol Neurobiol 49:88–102

Giuliani P, Ballerini P, Ciccarelli R, Buccella S, Romano S, D'Alimonte I, Poli A, Beraudi A, Peña E, Jiang S, Rathbone MP, Caciagli F, Di Iorio P (2012a) Tissue distribution and metabolism of guanosine in rats following intraperitoneal injection. J Biol Regul Homeost Agents 26:51–65

Giuliani P, Romano S, Ballerini P, Ciccarelli R, Petragnani N, Cicchitti S, Zuccarini M, Jiang S, Rathbone MP, Caciagli F, Di Iorio P (2012b) Protective activity of guanosine in an in vitro model of Parkinson's disease. Panminerva Med 54(1 Suppl 4):43–51

Häcker G (2000) The morphology of apoptosis. Cell Tissue Res 301:5–17

Han BH, Holtzman DM (2000) BDNF protects the neonatal brain from hypoxic-ischemic injury in vivo via the ERK pathway. J Neurosci 20:5775–5781

Hansson E, Sellstrom A (1983) MAO, COMT, and GABA-T activities in primary astroglial cultures. J Neurochem 40:220–225

Inazu M, Kubota N, Takeda H, Zhang J, Kiuchi Y, Oguchi K, Matsumiya T (1999a) Pharmacological characterization of dopamine transport in cultured rat astrocytes. Life Sci 64:2239–2245

Inazu M, Takeda H, Ikoshi H, Uchida Y, Kubota N, Kiuchi Y, Oguchi K, Matsumiya T (1999b) Regulation of dopamine uptake by basic fibroblast growth factor and epidermal growth factor in cultured rat astrocytes. Neurosci Res 34:235–244

Jin K, Mao XO, Zhu Y, Greenberg DA (2002) MEK and ERK protect hypoxic cortical neurons via phosphorylation of Bad. J Neurochem 80:119–125

Jin CM, Yang YJ, Huang HS, Kai M, Lee MK (2010) Mechanisms of L-DOPA-induced cytotoxicity in rat adrenal pheochromocytoma cells: implication of oxidative stress-related kinases and cyclic AMP. Neuroscience 170:390–398

Kulich SM, Chu CT (2001) Sustained extracellular signal-regulated kinase activation by 6-hydroxydopamine: implications for Parkinson's disease. J Neurochem 77:1058–1066

Lee C, Park GH, Jang JH (2011) Cellular antioxidant adaptive survival response to 6-hydroxydopamine-induced nitrosative cell death in C6 glioma cells. Toxicology 283:118–128

Leist M, Single B, Castoldi AF, Kuhnle S, Nicotera P (1997) Intracellular adenosine triphosphate (ATP) concentration: a switch in the decision between apoptosis and necrosis. J Exp Med 185:1481–1486

Makar TK, Nedergaard M, Preuss A, Gelbard AS, Perumal AS, Cooper AJL (1994) Vitamin E, ascorbate, glutathione, glutathione disulfide, and enzymes of glutathione metabolism in cultures of chick astrocytes and neurons: evidence that astrocytes play an important role in antioxidative processes in the brain. J Neurochem 62:45–53

Mayo JC, Sainz RM, Antolín I, Rodriguez C (1999) Ultrastructural confirmation of neuronal protection by melatonin against the neurotoxin 6-hydroxydopamine cell damage. Brain Res 818:221–227

McGeer PL, McGeer EG (2008) Glial reactions in Parkinson's disease. Mov Disord 23:474–483

McNaught KS, Perl DP, Brownell AL, Olanow CW (2004) Systemic exposure to proteasome inhibitors causes a progressive model of Parkinson's disease. Ann Neurol 56:149–162

Mena MA, García de Yébenes J (2008) Glial cells as players in parkinsonism: the 'good', the 'bad', and the 'mysterious' glia. Neuroscientist 14:544–560

Parkinson FE, Ferguson J, Zamzow CR, Xiong W (2006) Gene expression for enzymes and transporters involved in regulating adenosine and inosine levels in rat forebrain neurons, astrocytes and C6 glioma cells. J Neurosci Res 84:801–808

Pelton EW, Kimelberg HK, Shipherd SV, Bourke RS (1981) Dopamine and norepinephrine uptake and metabolism by astroglial cells in culture. Life Sci 28:1655–1663

Pettifer KM, Kleywegt S, Bau CJ, Ramsbottom JD, Vertes E, Ciccarelli R, Caciagli F, Werstiuk ES, Rathbone MP (2004) Guanosine protects SH-SY5Y cells against beta-amyloid-induced apoptosis. Neuroreport 15:833–836

Pettifer KM, Jiang S, Bau C, Ballerini P, D'Alimonte I, Werstiuk ES, Rathbone MP (2007) MPP(+)-induced cytotoxicity in neuroblastoma cells: antagonism and reversal by guanosine. Purinergic Signal 3:399–409

Sagara JI, Miura K, Bannai S (1993) Maintenance of neuronal glutathione by glial cells. J Neurochem 61:1672–1676

Su C, Elfeki N, Ballerini P, D'Alimonte I, Bau C, Ciccarelli R, Caciagli F, Gabriele J, Jiang S (2009) Guanosine improves motor behavior, reduces apoptosis, and stimulates neurogenesis in rats with parkinsonism. J Neurosci Res 87:617–625

Timmons S, Coakley MF, Moloney AM, O' Neill C (2009) Akt signal transduction dysfunction in Parkinson's disease. Neurosci Lett 467:30–35

Traversa U, Bombi G, Di Iorio P, Ciccarelli R, Werstiuk ES, Rathbone MP (2002) Specific [(3)H]-guanosine binding sites in rat brain membranes. Br J Pharmacol 135:969–976

Traversa U, Bombi G, Camaioni E, Macchiarulo A, Costantino G, Palmier C, Caciagli F, Pellicciari R (2003) Rat brain guanosine binding site. Biological studies and pseudo-receptor construction. Bioorg Med Chem 11:5417–5425

Advs Exp. Medicine, Biology - Neuroscience and Respiration (2015) 6: 35–39
DOI 10.1007/5584_2014_66

Published online: 14 October 2014

Chemoresponsiveness and Breath Physiology in Anosmia

Andrea Mazzatenta, Mieczyslaw Pokorski, Danilo Montinaro, and Camillo Di Giulio

Abstract

Anosmia is a model to study the interaction among chemoreception systems. In the head injury, the traumatic irreversible anosmia caused by damage to olfactory nerve fibers and brain regions is of enviable research interest. In this study, psychophysiological tests for a comprehensive assessment of olfactory function were utilized to investigate anosmia, together with a new technique based on the breath real-time monitoring of volatile organic compounds (VOCs). We applied the breath and VOCs analysis to investigate chemoresponsiveness in the long-term irreversible post-traumatic anosmia.

Keywords

Anosmia • Breath analysis • Continuous monitoring • Metal oxide semiconductor sensor • Volatile organic compounds

A. Mazzatenta (✉)
Sensorial Physiology Unit, Department of Neuroscience and Imaging, "G. d'Annunzio" University of Chieti–Pescara, 66013 Chieti, Italy

Faculty of Veterinary Medicine, University of Teramo, Teramo, Italy

Neuroscience Institute, C.N.R., Pisa, Italy
e-mail: amazzatenta@yahoo.com

M. Pokorski
Department of Nursing, Public Higher Medical Professional School in Opole, 68 Katowicka St., 45-060 Opole, Poland

Institute of Psychology, University of Opole, Opole, Poland
e-mail: m_pokorski@hotmail.com

D. Montinaro
Sensorial Physiology Unit, Department of Neuroscience and Imaging, "G. d'Annunzio" University of Chieti–Pescara, 66013 Chieti, Italy

Department of Psychiatry and Psychology, ASL 2, Chieti, Italy

1 Introduction

Smell is essential to life and it is a specialized chemoreceptive sensorial system. The investigation of olfaction and its cross-modal interaction with other chemoreceptor systems is rather a complicated matter. Anosmia could be a model for such a study. There are essentially two categories of anosmia. A constitutive anosmia, which is a disorder linked to a particular genetic alteration, e.g., the Kallmann syndrome. A traumatic anosmia, on the other hand, is induced by head injury, chemical exposition, pharmacological treatment, surgical damage, or psycho-

C. Di Giulio
Sensorial Physiology Unit, Department of Neuroscience and Imaging, "G. d'Annunzio" University of Chieti–Pescara, 66013 Chieti, Italy

logical stress. Head injury is the most common cause of smell impairment (Temmel et al. 2002; Sumner 1975; Sumner 1964). The neural mechanisms for post-traumatic anosmia are usually damage to the olfactory nerve fibers and/or brain regions (Doty et al. 1997; Costanzo and Zasler 1991). Dramatically, the severity of olfactory loss is not correlated with the prognosis (Zusho 1982). The recovery becomes less expectable when anosmia lasts over 1–2 year period, thus indicating an irreversible damage in the olfactory neural pathway (Mueller and Hummel 2009; Costanzo and Becker 1986; Sumner 1964). The clinical approach to diagnose the severity of anosmia is through the University of Pennsylvania Smell Identification test (UPSIT) (Doty et al. 1984), the 'Sniffin' Sticks' test (SS) (Hummel et al. 1997), and the Connecticut Chemosensory Clinical Research Center test (CCCRC) (Cain et al. 1988) which are the most commonly used psychophysiological tests for a comprehensive assessment of olfactory function. Clinical instrumental analysis is through the electrophysiological techniques: the olfactory electroencephalogram (O-EEG) and the olfactory event-related potentials (OERP) (Mueller and Hummel 2009). However, a new technique based on the breath real-time monitoring of volatile organic compounds (VOCs) has a potential to be useful in the study and diagnosis of anosmia (see Mazzatenta et al. 2013a for a review of VOCs analysis). The real-time monitoring of VOCs enables to assess both the effect of olfactory stressor and the level of central neural fatigue in both healthy and sick subjects in combination with cognitive performance (Mazzatenta et al. 2013b, c). Here, we applied the breath real-time monitoring of VOCs to investigate the chemo-responsiveness in the long-term irreversible post-traumatic anosmia.

2 Methods

2.1 Case Presentation

An anosmic Caucasian male patient, aged 62, volunteered for the study. The patient provided written informed consent. The procedure was performed in agreement with the Ethical Principles of the Helsinki Declaration and was accepted by a local Ethics Committee. The patient suffered from a complete irreversible anosmia for 32 years after a car accident in 1981, which caused a severe frontal and basal head trauma. The olfacto-electroencephalography (OEEG) and olfactory event-related potentials (OERP) were both negative. Patient's history suggested post-traumatic olfactory loss, possibly due to shearing of olfactory nerve fibers passing through the ethmoidal cribriform plate and contusion of the olfactory bulb and frontal lobe. Smell function as well as flavor perception during eating and drinking also were completely lost.

2.2 Protocol and Measurements

Preliminary psychophysiological tests consisting of olfactory threshold, discrimination, and identification were performed. The olfactory threshold was investigated by applying a standard n-butanol test (woody-alcohol odor), using a series of nine increasing molar concentrations ranging from 10^{-5} to 0.6 M in the following order: $9.14e^{-5}$, $2.74e^{-4}$, $8.23e^{-4}$, 0.00245, 0.0074, 0.022, 0.67, 0.2, and 0.6 M (Cain 1982). The trial ended when the subject perceived an olfactory sensation. The test was re-administered three times, and the time interval between stimuli was about 30 s. The interval between trials with increasing concentrations of the n-butanol solutions was at least 5 min. The subject was stimulated for 3–4 s at a distance of 2 cm from the nostrils. The threshold value was defined as the average score of dilution steps. Olfactory discrimination was evaluated by asking to find the odor source between three vials, one containing the stimulus (the stimuli used are: R(-)-carvone, cinnamaldehyde, citral, eugenol, and 1-octen-3-ol) and two without it. The olfactory identification was evaluated by using a forced test with four possible answers (one correct, one confounding, and two wrong), by using a series of 12 stimuli (acetophenone, R(−) and S(+) carvone, cinnamaldehyde, citral, citronellal, eucalyptol, eugenol, hexanal, isoamylacetate, 1-octen-3-ol, and phenethyl alcohol). The stimulation was for

3–4 s at 2 cm from the nostrils, the time interval between two stimuli was about 45 s.

After the investigation of olfaction, the real-time assessments of breath pattern and exhaled breath content in the control condition and after exposure to the olfactory stimulus, n-butanol were performed. The recording system used in this experiments was an iAQ-2000 (Applied Sensor, Warren, NJ) equipped with a metal oxide semiconductor (MOS) having a sensing range of 450–2,000 ppm CO_2 equivalents which is able to detect a broad range of volatile compounds (both organic and inorganic, e.g., alcohols, aldehydes, aliphatic hydrocarbons, amines, aromatic hydrocarbons, ketones, organic acids and CO), while correlating directly with the CO_2 levels (Mazzatenta et al. 2013b, c).

The breath and VOCs signals were calculated as an integral of the signal curve in a time interval. The normalization applied was according to the following formula:

$$X = \left(x_i - x_{i,j}\min\right) / \left(x_{i,j}\max - x_{i,j}\min\right)$$

where X is the normalized value obtained from the result of subtraction from a value x the minimal value in a series of values ranging from *i* to *j* divided by the result of subtraction from the maximal value in the same series the minimal value. An example of the normalization formula in a series of values of 54, 23, and 68 is this: (54 − 23)/(68 − 23) = 31/45 = 0.69; the 23 becomes 0; and 68 becomes 1.

Data were given as means ± SD. Differences between pre- and post-stimulatory conditions were assessed with one-way ANOVA. The α-level was set at 0.05. Statistical analysis was done with Excel, Origin, and SPSS software.

3 Results

The psychophysiological tests performed showed severe anosmia to n-butanol; the olfactory threshold was greater than 0.6 M. Accordingly, olfactory discrimination and identification investigation resulted in a null score.

The real-time recordings of the breath signal showed appreciable alterations in breathing after 3 s of 0.6 M n-butanol stimulation, consisting of increased both minimal and maximal breath amplitude (Fig. 1); the increases abated down to the pre-stimulus level within the next 30 s. Quantification of the normalized minimal and maximal breath amplitudes is shown in Fig. 2a, b, respectively. Significance of the pre-post n-butanol increase was confirmed for the peak breath amplitude only (Fig. 2b) (one-way ANOVA $F_{(1,76)} = 15.92$; $p < 0.001$). Likewise, separate analysis of the minimal-maximal breath amplitude differences in the pre-n-butanol (control) condition and the post-n-butanol condition showed a significance increase only for the post-n-butanol condition (one-way ANOVA $F_{(1,65)} = 20.3$; $p < 0.001$).

The real-time monitoring of exhaled breath content showed a rapid increase in VOCs exhaled in the post-n-butanol phase; the stimulation with 0.6 M n-butanol took 3 s and a significant increase in VOCs appeared about 30 s after the stimulation (Fig. 3) (one way ANOVA $F_{(1,168)} = 17.65$; $p < 0.001$).

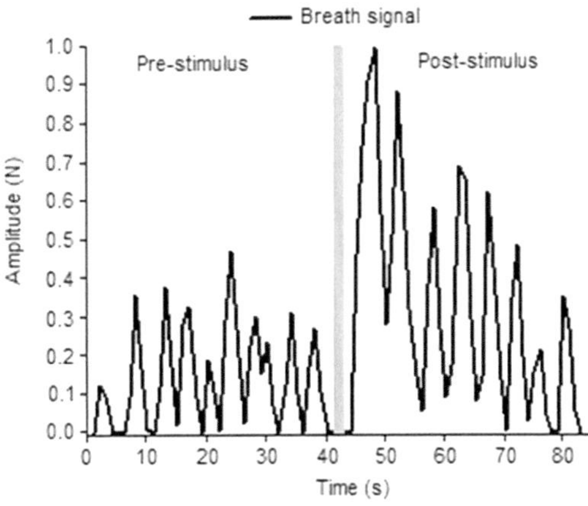

Fig. 1 Breath real-time monitoring in a post-traumatic anosmic subject. Breath signal amplitude increased immediately after 3 s of stimulation with 0.6 M n-butanol. *Gray bar* represents the onset of stimulation

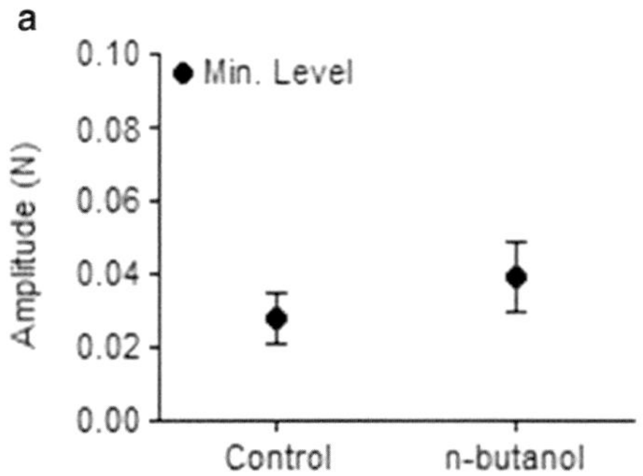

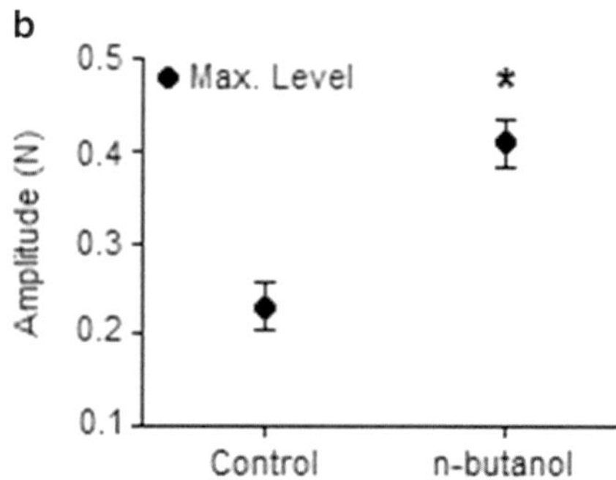

Fig. 2 **Comparison of the minimal (a) and maximal (b) levels of peak breath amplitudes** in control *vs.* after 3 s of stimulation with 0.6 M n-butanol. Significant difference concerns only the maximal levels of peak amplitude ($p < 0.001$). Note different scales in each panel

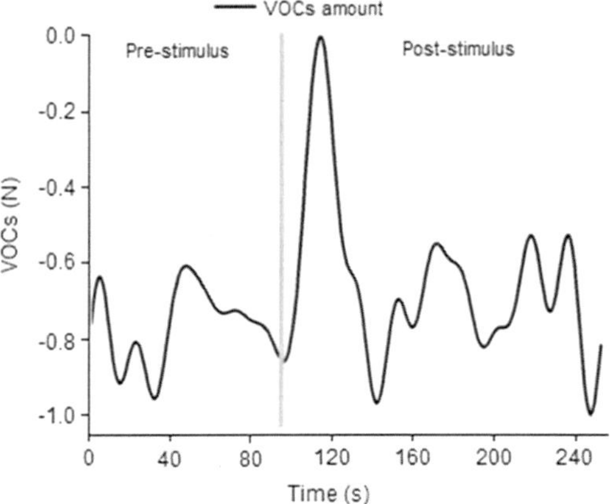

Fig. 3 Breath real-time monitoring of exhaled VOCs before and after stimulation with 0.6 M n-butanol. An appreciable increase in VOCs appeared about 30 s post-stimulation

4 Discussion

Recovery from post-traumatic olfactory loss is reported in the literature to be barely about 10 % (Reden et al. 2006; Zusho 1982). In most cases, it takes a few months until first olfactory impressions are reported, in some few cases it may take up to several years (Mueller and Hummel 2009; Sumner 1964). The patient studied in the present paper was a permanently anosmic subject whit a negative response to electrophysiological and neuro-psychophysiological assessments; his olfactory threshold was greater than 0.6 M n-butanol, the maximum concentration used in the test, representing the first dilution step. The real-time recording of breath signal showed a response to n-butanol, besides the subject no difference distinguish between pre- nor in stimulation nor in post-stimulation. However, the breath response, consisting of increased maximal breath amplitude, was analogous to the stress response previously reported in normosmic subjects (Mazzatenta et al. 2013b). This interpretation is corroborated by a significant change in the average maximum breath peak, while not in the minimum one; the latter suggests a regular basal physiological activity. Furthermore, the amount of VOCs increased dramatically following the n-butanol administration, which is clearly related to an alteration in metabolism. Such metabolic alterations could be explained by environmental stressors (Mazzatenta et al. 2013b) or neural fatigue (Mazzatenta et al. 2013c); in either case chemical information reaches the homeostatic regulatory center by a pathway distinctly different from the olfactory chemoreceptive one.

In the context of the study model of permanent complete anosmia, the disparity between olfactory and breathing results raises interest in that it gives a new insight into the complexity of chemoreception systems. The explanation of this disparity could be a cross-modal interaction between several system of chemoreception, e.g., olfactory, carotid body chemoreceptors, and trigeminal ones. When one element of the system is missing, others sustain their activity. However, it also is reasonable to speculate on the presence of a phenomenon of neural plasticity based on adaptive mechanisms. When the olfactory system is severely compromised, other chemoreceptive systems could be recruited to substitute for its target action.

We therefore believe the results presented in this study may stimulate investigations into the chemoreception considered as a subtle interaction between different, and up to now thought of as being distinctly independent, sensorial systems. This interaction seems underlain by adaptive neural mechanisms addressing the detection of percepts in the surrounding environment.

Conflicts of Interest The authors declare no conflicts of interest in relation to this article.

References

Cain WS (1982) Sumner's 'On testing the sense of smell' revisited. Yale J Biol Med 55:515–519

Cain WS, Gent JP, Goodspeed RD, Leonard G (1988) Evaluation of olfactory dysfunction in the Connecticut Chemosensory Clinical Research Center. Laryngoscope 98:83–88

Costanzo RM, Becker DP (1986) Smell and taste disorders in head injury and neurosurgery patients. In: Meiselman HL, Rivlin RS (eds) Clinical measurements of taste and smell. MacMillan Publishing Company, New York, pp 565–578

Costanzo RM, Zasler ND (1991) Head trauma. In: Getchell TV, Doty RL, Bartoshuk LM, Snow JB Jr (eds) Smell and taste in health and disease. Raven, New York, pp 711–730

Doty RL, Shaman P, Dunn M (1984) Development of the University of Pennsylvania Smell Identification Test: a standardized microencapsulated test of olfactory function. Physiol Behav 32:489–502

Doty RL, Yousem DM, Pham LT, Kreshak AA, Geckle R, Lee WW (1997) Olfactory dysfunction in patients with head trauma. Arch Neurol 54:1131–1145

Hummel T, Sekinger B, Wolf SR, Pauli E, Kobal G (1997) Sniffin' sticks: olfactory performance assessed by the combined testing of odor identification, odor discrimination and odor threshold. Chem Senses 22:39–52

Mazzatenta A, Di Giulio C, Pokorski M (2013a) Pathologies currently identified by exhaled biomarkers. Respir Physiol Neurobiol 187:128–134

Mazzatenta A, Pokorski M, Cozzutto S, Barbieri P, Veratti V, Di Giulio C (2013b) Non-invasive assessment of exhaled breath pattern in patients with multiple chemical sensibility disorder. Adv Exp Med Biol 756:179–188

Mazzatenta A, Pokorski M, Di Giulio C (2013c) Real time breath analysis in type 2 diabetes mellitus patients during cognitive task. Adv Exp Med Biol 788:247–253

Mueller CA, Hummel T (2009) Recovery of olfactory function after nine years of post-traumatic anosmia: a case report. J Med Case Rep 1–3:9283

Reden J, Mueller A, Mueller C, Konstantinidis I, Frasnelli J, Landis BN, Hummel T (2006) Recovery of olfactory function following closed head injury or infections of the upper respiratory tract. Arch Otolaryngol Head Neck Surg 132:265–269

Sumner D (1975) Disturbance of the senses of smell and taste after head injuries. In: Vinken PH, Bruyn GW (eds) Handbook of clinical neurology. North-Holland Publishing Company, Amsterdam, pp 1–25

Sumner D (1964) Post-traumatic anosmia. Brain 87:107–129

Temmel AF, Quint C, Schickinger-Fischer B, Klimek L, Stoller E, Hummel T (2002) Characteristics of olfactory disorders in relation to major causes of olfactory loss. Arch Otolaryngol Head Neck Surg 128:635–641

Zusho H (1982) Posttraumatic anosmia. Arch Otolaryngol 108:90–92

Advs Exp. Medicine, Biology - Neuroscience and Respiration (2015) 6: 41–47
DOI 10.1007/5584_2014_69

Published online: 14 October 2014

Cognitive Functioning of the Prelingually Deaf Adults

Mieczysław Pokorski and Sandra Klimańska

Abstract

Deafness is a model of brain adaptation to sensory deprivation which entails psychomotor and cognitive domains. This study seeks to determine the level of emotional intelligence, assessed from the ability to discern emotions from facial expressions, visual and mental attention, and non-verbal fluency in the deaf people as compared with the hearing counterparts. Participants were 29 prelingually deaf, hearing loss of >70 dB, communicating only in sign language, and 30 hearing persons. The age range of all subjects was 40–50 years. Psychometric tools consisted of the Emotional Intelligence Scale-Faces, the d2 Test of Attention, and the Figural Fluency Test. Data elaboration took gender into account. The findings were that both deaf women and men defined significantly fewer emotions as known, compared with the hearing persons. However, the deaf men, but not women, were able to properly recognize a higher percentage of emotions associated with a definite face look, among the emotions they knew. There were no appreciable differences in attention indices between the deaf and hearing men, but deaf women's total performance on attention was worse. By contrast, deaf women, but not men, fared better in non-verbal fluency, compared with their hearing counterparts. We conclude that, on the whole, prelingual deafness does not impede cognitive functioning in adult age. The nature of detecting and executing of cognitive tasks, despite gender and task-specific variations, is preserved. Brain networks are able to compensate for the missing auditory input.

Keywords

Attention • Cognitive functioning • Emotional intelligence • Nonverbal fluency • Prelingual deafness

M. Pokorski (✉)
Department of Nursing, Public Higher Medical Professional School in Opole, 68 Katowicka St., 45-060 Opole, Poland
e-mail: m_pokorski@hotmail.com

S. Klimańska
Institute of Psychology, Opole University, Opole, Poland

1 Introduction

Prelingual deafness is a model of brain functional reorganization to develop adaptive processes mitigating sensory deprivation. Various cognitive abilities among the deaf have been studied in the past, but the vast majority of the researched groups consisted of children or adolescents at school age (Oberg and Lukomski 2011). The majority of studies show that cognitive functioning in deaf children is, in the main, on par with normal children, but some studies underline differences in scholarly achievements likely caused by parental or environmental factors (Lauwerier et al. 2003). Brain neuronal networks specialized in auditory processing are subject to functional plasticity long after the childhood physiological brain maturation (Lazard et al. 2013). The effects of environmental factors seem enhanced in adulthood, as hearing disability is often treated as handicap with all the negative psychological connotations involving stigmatization, difficulties in career development and personal advances, often resulting in lower social status compared with the hearing persons. Environmental stressors have an impact on cognition (Brosschot et al. 2006) and thus are viable to affect plastic brain networks' reorganization to offset auditory deprivation. On the advantage side, the prelingually deaf rely on the visual communication skills such as sign language or face emotion recognition, which implements interaction between brain sensory and cognitive domains. Eyesight is the main sensory connector to the surrounding environment, and is some kind of sensory compensation for the missing hearing sense in the deaf. The delayed adulthood face of emotional and cognitive functioning of prelingually deaf persons is thus less certain.

The present study was designed to examine the emotion-processing skills, assessed from the ability to discern emotions from facial expressions, visual and mental attention, and non-verbal fluency in the prelingually deaf people as compared with the hearing counterparts.

2 Methods

The study was approved by a Review Board for Research of the Institute of Psychology of Opole University in Poland and was performed in accord with the Declaration of Helsinki for Human Research. All study participants gave informed consent concerning the survey procedures employed. A total of 59 subjects were enrolled into the study. There were two groups: 29 persons with hearing loss (F/M – 15/14) and 30 hearing persons (F/M – 15/15). The age of all subjects was within the 40–50 years range. Deaf persons had a severe hearing loss of greater than 70 dB since early childhood and could communicate only in sign language. Those with normal hearing decelerated that they never had any hearing problems and were unable to communicate in sign language. The inclusion criteria included a generally good health status, no history of major head injury or brain diseases, no current medical or psychological problems, no overt sensory impairments, and no prescription medication use.

Each person of both groups completed a battery of the following tests: the Emotional Intelligence Scale – Faces (EIS-F) developed and verified for the Polish population by Matczak et al. (2005), the d2 Test of Attention of Brickenkamp and Zillmer (1998), and the Figural Fluency Test (RFFT) of Ruff (1996). The EIS-F scale assesses the ability to discern emotions on the basis of facial expression. Mimic expression of emotions is considered an essential feature of emotional intelligence. The test consists of 18 sets of photos, each set shows six differently named emotions and the person should match the name with emotion. Since some words may be abstract to the deaf who never hear them, and who thus may have problems to connect a written word to the emotion seen in the picture or cannot describe it in one word, the test instruction was modified in an effort to equalize the chances of correct response for the deaf and hearing persons. The original instruction saying: if you are unsure whether a given face presents this concrete emotion, choose 'hard to say' was modified to: 'if you are unsure whether a given face presents this concrete emotion *or you do not know the name of emotion*, choose 'hard to say''.

The d2 Test of Attention assesses individual psychomotor performance, consisting of visual and mental attention, concentration, and processing speed and quality (Zillmer and

Kennedy 1999). The test is composed of the letters *d* and *p* with a variable number of dashes underneath or above the letter, from two to four, arranged in a random order. The subject is asked to find out *d's* with two dashes only (correct hits). The letters are arranged in 14 lines of 47 characters each, and the subject has a limited time of 20 s per line. Corrections are permitted. Performance is characterized by the total number of all letters processed (TN), the number of errors of omission (E1), the number of wrong letters crossed out (E2), the total number of errors (E = E1 + E2), the percentage of errors (E%), total performance (TP = TN−E), fluctuation rate (FR), i.e., the maximum difference between TN across individual trials, and the index of concentration performance (CP), i.e., the number of correctly processed letters minus E (see details in Table 3).

The RFFT is also a timed test which measures the ability to generate as many new figures (unique designs) as possible by connecting at least two dots included in a 5-dot matrix in such a way that each time a new figure would be created. The test consists of five parts with some variations in pattern design, except for parts II and III which contain the same pattern of dots, but with different distracting elements of the type of diamonds or lines. The time limit per part was set for 1 min. Repetition of the figures should be avoided as much as possible. This measure of non-verbal fluency takes into account the number of unique designs and perseverations a person makes (Ruff et al. 1987).

Surveys were conducted during individual sessions under the oversight of a test supervisor. Completion of all questionnaires, on average, required 45 min. Data were presented as means ± SD of raw scores. The normality of data distribution was determined with Kolmogorov-Smirnov's test. Equality of variances for a variable calculated for two groups was assessed with Lavene's test. In case of unequal variances the corresponding variables were compared with Mann-Whitney U test. Otherwise, statistical analysis consisted of an unpaired or paired *t*-test, as required, for two group comparisons. The level of statistical significance was set at $p < 0.05$. A commercial statistical package SPSS ver. 19 was used for the statistical elaboration.

3 Results

3.1 Emotional Intelligence – Recognition of Face Emotions

Analysis of the gender-combined groups of the deaf ($n = 29$) and hearing persons ($n = 30$) demonstrated appreciable differences in the number of known and unknown face emotions being seen in the pictures. The deaf correctly identified known emotions in 33.1 ± 12.2 female and 33.4 ± 12.4 male faces, whereas unknown emotions were reported in 20.3 ± 12.3 and 20.4 ± 12.5 faces, respectively. The number of correct hits in the hearing persons was 52.8 ± 1.6 and 53.2 ± 1.0 for the known emotions in female and male faces, whereas the unknown emotions were reported in 0.9 ± 1.4 and 0.6 ± 0.9 faces, respectively. All differences between the deaf and hearing persons in the recognition of the faces of the corresponding gender being looked upon were significant ($p = 0.001$). However, there were no significant differences between the number of known or unknown female and male faces recognized by the deaf or hearing persons.

Since the perception of face emotions could differ not only regarding the gender of the faces being observed, but also of the observer's gender we broke down the data by the observers' gender for further analysis. The data on the number of female and male face emotions, defined as known or unknown, identified by the deaf and hearing women and men are shown in Table 1. Both deaf women and men defined significantly fewer emotions as known, and thus more as unknown, compared with the hearing counterparts. That held true for emotions seen in both female and male faces. These quantitative results clearly show that the deaf were not able to name all emotions that have a mimic representation.

The number of identified emotions is not tantamount to the ability of emotion recognition. Therefore, the results were subjected to further

Table 1 Number of emotions seen in pictures of female and male faces recognized and unrecognized by deaf and hearing persons in a test for emotional intelligence

	Women		Men	
	Deaf (*n* = *15*)	Hearing (*n* = *15*)	Deaf (*n* = *14*)	Hearing (*n* = *15*)
Female face emotions				
Recognized	34.9 ± 11.2*	53.0 ± 1.2†	31.2 ± 13.3*	52.6 ± 1.9†
Unrecognized	18.4 ± 11.6*	0.8 ± 1.1	22.4 ± 13.1*	1.0 ± 1.6
Male face emotions				
Recognized	30.6 ± 11.1*	52.9 ± 1.2†	30.6 ± 13.6*	53.5 ± 0.6†
Unrecognized	17.9 ± 11.1*	0.9 ± 1.0	20.3 ± 13.1*	0.3 ± 0.5

Values are means ± SD of cases
*$p \leq 0.001$ for differences between deaf and hearing persons within a gender (horizontal pairwise comparison)
†$p \leq 0.01$ for differences between recognized and unrecognized emotions within a gender (vertical pairwise comparison)

Table 2 Emotions recognized in female and male faces by deaf and hearing persons as percentage of all known emotions in a given group

	Women		Men	
	Deaf (*n* = *15*)	Hearing (*n* = *15*)	Deaf (*n* = *14*)	Hearing (*n* = *15*)
Female face emotions	59.1 ± 12.3	55.5 ± 8.3	66.8 ± 10.3*	54.8 ± 11.7
Male face emotions	55.1 ± 13.6	56.9 ± 8.5	63.6 ± 6.5*	53.5 ± 7.7

Values are means ± SD of percent
*$p \leq 0.01$ for differences between deaf and hearing persons within a gender

qualitative analysis consisting of the calculation of a proportion of correct hits of face emotions among all known emotions. The results of this calculation broke down by the gender of the deaf and hearing observers of faces and the gender of the faces being looked upon are shown in Table 2. The deaf men were able to properly recognize more emotions associated with a definite either female or male face look, among the emotions they knew, compared with the hearing counterparts. That regularity did not concern the deaf women. Thus, it might be presumed that the deaf men are able to use the visuospatial coordination more effectively and readily in performance, to the extent it partakes of the emotional intelligence.

3.2 Attention – d2 Test

The test assessed the ability to scan visual material by comparing the number of mistakes made along the process in the deaf and hearing persons. There were inter-gender differences in the assessment of attention. The deaf women, but not men, crossed out significantly fewer letters than the hearing ones ($p < 0.05$), whereas the number of errors made was akin irrespective of the hearing status (Table 3). It follows that the deaf women's total performance corrected for the number of errors made and concentration was worse. On the positive side, the scatter of the total number of letters processed in individual trials was significantly smaller in the deaf than that in the hearing women ($p < 0.05$). In contrast, differences in the attention indices between the deaf and hearing men were inappreciable. The hearing men, however, performed overall worse on the attention score than the hearing women in terms of the total and error-corrected numbers of letters processed and concentration. This inter-gender difference was absent in the deaf subjects whose attention was on a par in both genders.

3.3 Non-verbal Fluency: Ruff Figural Test (RFFT)

The RFFT was used to compare the non-verbal fluency between the deaf and hearing persons. Quantitative results were expressed as the total

Table 3 Comparative descriptions of variables in the d2 attention test in the deaf and hearing persons stratified by gender

	Women		Men	
	Deaf (*n* = *15*)	Hearing (*n* = *15*)	Deaf (*n* = *14*)	Hearing (*n* = *15*)
TN	420.3 ± 91.9*	483.5 ± 81.7	407.9 ± 99.8	404.2 ± 63.7†
E1	33.3 ± 24.2	32.6 ± 18.3	23.4 ± 19.8	30.1 ± 23.3
E2	7.1 ± 4.9	8.9 ± 6.0	9.9 ± 8.0	6.5 ± 5.1
E	40.3 ± 24.7	41.5 ± 19.0	33.3 ± 25.6	36.6 ± 25.8
E%	9.6 ± 5.1	8.7 ± 4.1	7.6 ± 4.7	8.7 ± 5.6
TP	380.2 ± 84.9*	442.1 ± 83.0	374.6 ± 84.4	367.6 ± 51.4†
FR	19.1 ± 6.7*	13.1 ± 5.3	19.1 ± 7.5	20.1 ± 6.0†
CP	143.5 ± 35.1	166.9 ± 52.4	142.8 ± 26.0*	120.5 ± 25.9†

Values are means ± SD. *TN* total number of all letters processed, *E1* omissions, *E2* wrong letters crossed out, *E* total number of errors (E1 + E2), *E%* percentage of errors, *TP* total performance = TN-E; FR-fluctuation rate, i.e., the maximum difference between TN across individual trials, *CP-index of concentration performance* – the number of correctly processed letters minus E

*p ≤ 0.05 for differences between deaf and hearing persons within a gender

†p ≤ 0.05 for inter-gender differences within the same hearing state

Table 4 Non-verbal fluency (RFF Test) in deaf and hearing persons stratified by gender

	Women		Men	
	Deaf (*n* = *15*)	Hearing (*n* = *15*)	Deaf (*n* = *14*)	Hearing (*n* = *15*)
Unique designs				
Without distractors	63.5 ± 11.4	68.1 ± 15.7	56.1 ± 19.2	60.3 ± 15.8
With distractors	42.5 ± 9.2	43.7 ± 11.1	39.9 ± 15.9	41.2 ± 10.6
Perseverations				
Without distractors	1.5 ± 1.9	3.5 ± 2.6*	5.5 ± 7.5	3.5 ± 2.4
With distractors	1.9 ± 2.7	2.3 ± 2.1	1.2 ± 1.7	1.1 ± 1.1
Error rate (%)				
Without distractors	2.2 ± 2.7	5.5 ± 4.6*	10.4 ± 17.6	5.3 ± 3.0
With distractors	3.6 ± 5.1	5.2 ± 4.4	2.9 ± 3.4	2.9 ± 2.6

Values are means ± SD

*p ≤ 0.05 for differences between deaf and hearing persons within a gender (horizontal pairwise comparison)

number of unique designs created and the total number of perseverative errors, whereas qualitative results were expressed as the percentage rate estimated from the ratio of perseverative errors to the total number of unique designs. Performance on the RFFT was measured without and with distractors.

Overall, the deaf tended to generate fewer unique designs, which was slightly more accentuated in the part of the test without distractors, than the hearing persons. There was also an inter-gender tendency observed in the number of unique designs toward the disadvantage of men regarding the number of unique designs generated, irrespective of deafness and the presence or not of distractors (Table 4). The deaf women made significantly fewer repetitions of already generated figures and the proportion of perseverations in the total number of figures generated was smaller than that in their hearing counterparts; the findings were absent in the deaf men. The RFFT results demonstrate a slightly better non-verbal fluency in the deaf women, but not men, compared with their hearing counterparts, although the differences were rather small and the deaf women tended to generate fewer unique designs.

4 Discussion

The present study contrasted the emotion-processing skills, visuospatial and mental attention, and non-verbal fluency in the prelingually deaf and normal hearing persons. Prelingual deafness is hearing damage that arises before the acquisition of language, which has a congenital background or arises in early infancy due to morbid conditions (Rehm et al. 2003). This kind of deafness impairs the individual's ability to acquire sound of spoken language. Prelingual deafness entails psychomotor and cognitive changes, which may be exemplified by the development of visual communication skills, such as sign language or recognition of facial expressions. Sign language transmits meaning by manual and body communication instead of acoustically transmitted sound. This involves a mixture of fine movements of hands and arms, accompanied by face emotions. The deaf persons' fluid movements serve to express aspects of meaning. It could thus be expected that prelingual deafness would particularly engage visuospatial and attentional aspects of neural connections.

The findings of the present study demonstrate that deaf men were able to properly recognize a higher percentage of emotions associated with a definite face look, among the emotions they knew, compared with the hearing men. That result speaks in favor of enhanced visuospatial skills and emotional intelligence in prelingual deafness, to the extent that the recognition of face emotions is an essential element of emotional intelligence (Meyer et al. 1990; Salovey and Mayer 1989). However, visuospatial dexterity concerned deaf men, but not women, and the representatives of both genders defined significantly fewer emotions as known, compared with the hearing persons. There were no appreciable differences in attention indices between the deaf and hearing men. On the other side, deaf women's total performance on attention was worse than that in the hearing women, although the deaf ones responded in a more regular and uniform way in attention trials. By contrast, deaf women, but not men, fared clearly better in non-verbal fluency, compared with their hearing counterparts. However, deaf women had difficulty in warding off distractors, as the edge they had over hearing women in non-verbal fluency clearly tended to wane in the presence of distractors.

Overall, therefore, we found that the brain deprived of auditory stimuli perception is able to adapt to remaining sensory inputs. The adaptive processes consolidated over time in the middle-aged individuals whose deafness appeared at the prelingual stage could lead to sustained alterations in the psychological domains. The alterations are, however, rather meager, mixed, and do not downgrade perceptual skills in the deaf. Conversely, functional brain reorganization that follows prelingual deafness alters one's interaction with the surrounding environment in a way that clearly tends to compensate for the missing sensory input. The deaf can more effectively concentrate on visual stimuli, which makes them notice more detail in visual material than the hearing persons do. That is in line with recent neurophysiological studies showing that people who are deaf benefit from a differently developed retina that enables to see better into the visual periphery compared with hearing adults, as shown in a non-invasive ocular coherence tomography that scans the retina (Codina et al. 2011). More neurons transmitting and processing somatosensory and visual stimulation have been shown in the cortex of deafened mice and the size of cortical auditory A1 area is increased (Hunt et al. 2006).

In the present study we found gender differences in visuospatial coordination in that it was better developed in deaf men than women. These differences seem an extension of those reported in visuospatial processing present in hearing adults. Sex differences in the rate of visuospatial processing and intelligence, favoring men, have been noted in behavioral and functional imaging studies (Weiss et al. 2003). Although right hemisphere predominates in visuospatial coordination in either sex, there are intra-hemisphere differences, with men showing parietal activation and women, in addition, inferior frontal activation during visuospatial tasks (Hugdahl et al. 2006). In this study, however, we failed to note appreciable advantage of hearing men over hearing women in visual aspects of face emotions recognition. We did note that hearing men and women use different processing strategies, favoring women, for attention-concentration and non-verbal fluency; the categorical representations ascribed to the left hemisphere (Rybash and Hoyer 1992). This advantage for attention processing was gone in deaf women, but persisted for non-verbal fluency. It seems, therefore, that the different processing strategies in men and women to solve similar tasks may have to do with hemisphere asymmetry, with right-sided bias for visuospatial tasks in men and left-sided bias for categorical tasks in women. The sex-related brain asymmetry seems little affected by the altered sensory experience of early onset deafness.

In conclusion, prelingual auditory deprivation promotes reorganization of brain networks that facilitates the input of signals related to visual communication and non-verbal fluency. The nature of detecting and executing of cognitive tasks, despite gender and task-specific variations, is preserved. We believe the study sheds complimentary light on the brain networks' ability to compensate for the missing auditory input, which allows the sensory-deprived individuals to function well.

Conflicts of Interest The authors declare no conflicts of interest in relation to this article.

References

Brickenkamp R, Zillmer E (1998) d2 test of attention. Hogrefe & Huber, Göttingen

Brosschot JF, Gerin W, Julian F, Thayer JF (2006) The perseverative cognition hypothesis: a review of worry, prolonged stress-related physiological activation, and health. J Psychosom Res 60:113–124

Codina C, Pascalis O, Mody C, Toomey P, Rose J, Gummer L, Buckley D (2011) Visual advantage in deaf adults linked to retinal changes. PLoS One 6: e20417. doi:10.1371/journal.pone.0020417

Hugdahl K, Thomsen T, Ersland L (2006) Sex differences in visuo-spatial processing: an fMRI study of mental rotation. Neuropsychologia 44:1575–1583

Hunt DL, Yamoah EN, Krubitzer L (2006) Multisensory plasticity in congenitally deaf mice: how are cortical areas functionally specified? Neuroscience 139:1507–1524

Lauwerier L, de Chouly de Lenclave MB, Bailly D (2003) Hearing impairment and cognitive development. Arch Pediatr 10:140–146 (Article in French)

Lazard DS, Innes-Brown H, Barone P (2013) Adaptation of the communicative brain to post-lingual deafness. Evidence from functional imaging. Hear Res 307:136–143

Matczak, M, Piekarska, J, and Studniarek, E (2005) SIE-T. Skala Inteligencji Emocjonalnej – Twarze. Psychological Test Laboratory of the Polish Psychological Association

Mayer JD, Maria DiPaolo M, Salovey P (1990) Perceiving affective content in ambiguous visual stimuli: a component of emotional intelligence. J Pers Assess 54:772–781

Oberg E, Lukomski J (2011) Executive functioning and the impact of a hearing loss: performance-based measures and the Behavior Rating Inventory of Executive Function (BRIEF). Child Neuropsychol: J Norm Abnorm Dev Child Adolesc 17:521–545

Rehm HL, Williamson RE, Kenna MA, Corey DP, Korf BR (2003) Understanding the genetics of deafness: a guide for patients and families. Harvard Medical School Center for Hereditary Deafness, Cambridge, MA

Ruff RM (1996) Ruff figural fluency test: professional manual. Psychological Assessment Resources, Lutz

Ruff RM, Light R, Evans R (1987) The ruff figural fluency test: a normative study with adults. Dev Neuropsychol 3:37–51

Rybash JM, Hoyer WJ (1992) Hemispheric specialization for categorical and coordinate spatial representations: a reappraisal. Mem Cogn 20:271–276

Salovey P, Mayer JD (1989) Emotional intelligence. Imagin Cogn Personal 9:185–211

Weiss E, Siedentopf CM, Hofer A, Diesenhammer EA, Hoptman MJ, Kremser C, Goalszewski S, Felber S, Fleischhacker WW, Delazer M (2003) Sex differences in brain activation pattern during a visuospatial cognitive task: a functional magnetic resonance imaging study in healthy volunteers. Neurosci Lett 344:169–1

Zillmer EA, Kennedy CH (1999) Construct validity for the d2 test of attention. Arch Clin Neuropsychol 14:728–735

Advs Exp. Medicine, Biology - Neuroscience and Respiration (2015) 6: 49–56
DOI 10.1007/5584_2014_84

Published online: 15 October 2014

Hypoxia-Related Brain Dysfunction in Forensic Medicine

R. Suslo, J. Trnka, J. Siewiera, and J. Drobnik

Abstract

Blood gases levels imbalances belong to important factors triggering central nervous system (CNS) functional disturbances. Hypoxia can be illness-related, like in many COPD patients, or it may be caused by broad range of external or iatrogenic factors – including influence of drugs depressing respiration, failure to keep the patient's prosthesis-supported airways patent, or a mistake in the operation of medical equipment supporting patient's respiration. Hypoxia, especially when it is not accompanied by rapid carbon dioxide retention, can go unnoticed for prolonged times, deepening existing CNS disorders, sometimes rapidly triggering their manifestation, or evoking quite new conditions and symptoms – like anxiety, agitation, aggressive behavior, euphoria, or hallucinations. Those, in turn, often result in situations raising interest in law enforcement institutions which need forensic medicine specialist's assistance and opinion. The possibility of illness or drug-related hypoxia, especially in terminal patients, is used to raise questions about the patients' ability to properly express their will in the way demanded by law – it also must be considered as a factor limiting the patients' responsibility in case they commit crimes. The possibility of hallucinations in hypoxia patients limits their credibility as witnesses or even their ability to report crime or sexual abuse they have been subjected to.

R. Suslo (✉)
Department of Forensic Medicine, Wroclaw Medical University, 4 Mikulicz-Radecki St., 50-345 Wroclaw, Poland

Public Health Department, University Hospital, Wroclaw, Poland
e-mail: robertsuslo@gmail.com

J. Trnka and J. Siewiera
Department of Forensic Medicine, Wroclaw Medical University, 4 Mikulicz-Radecki St., 50-345 Wroclaw, Poland

J. Drobnik
Department of Family Medicine, Wroclaw Medical University, Wroclaw, Poland

Department of Gynecology, Medical Profession High School, Opole, Poland

Keywords
Anesthesiology complication • Central nervous system • Court witness • Credibility system • Medical law • Oxygen depletion

1 Introduction

Hypoxia is defined in medical literature as an illness-related inadequate level of tissues oxygenation. It causes impairment of physiological functions, which threatens patient's health and life – especially in case of deep or rapidly progressing oxygen deficit. In the last decades many aspects concerning clinical hypoxia and results of this phenomenon have been discussed – including its etiology, mechanisms of arising, symptoms, clinical classifications, and methods of treatment – but both clinical medicine and forensic literature lack scientific reports concerning the importance of the generalized hypoxia, resulting from impaired gas exchange, in solving medical law-related problems. Hypoxia constitutes an important factor influencing both overall clinical patient's condition and the future prognosis; it also determines the impairment of different social functions. It must be stressed that generalized hypoxia-related central nervous system dysfunction impairs the patient's ability to make independent decisions as well as to declare the will in a lawful way. This, in turn, is being considered as the basis of patient's limited credibility. It constitutes quite a big practical problem to recognize the threshold beyond which the patient is unable to make independent informed decisions – including also the consent to undergo medical diagnosis and treatment. Chronic generalized hypoxia – by increasing anaerobic metabolism, and blood acid-alkali imbalances secondary to it – causes limitations of several mental functions and also behavioral changes that are developing silently and their manifestations can be often disguised by overlapping conditions and iatrogenic factors, especially drugs' side effects. It can result in delayed mental state alterations and seemingly unpredictable behaviors. Many different aspects of chronic generalized hypoxia – including its medical, social, and law-related aspects – shall influence the forensic medicine specialists' approach to cases when medico-legal opinion is required by law-enforcement institutions, penal, and civil courts.

2 Methods

The study has been approved by a local Ethics Committee. The aim of the study was to examine the chronic generalized hypoxia as an important factor considered when medico-legal opinions were required, mainly by law-enforcement institutions. The study was based on personal experience of the authors, documented in medico-legal opinions issued, supported by information acquired by an extensive review of the available literature and publications pertaining to the problem. The medico-legal archive cases issued by the teams of experts from the Department of Forensic Medicine of Medical University in Wroclaw, Poland in the years 2002–2012 have been reviewed to find cases in which chronic generalized hypoxia could have an impact on a patient's mental status and had been identified as the factor influencing the opinion's conclusion, or the possibility of hypoxia was raised in the questions the experts' team had been asked. The discussion of the cases chosen was enriched by comments based on the current state of medical and medico-legal knowledge.

3 Results

Three cases were selected from the archived medico-legal opinions, in which the chronic generalized hypoxia differential diagnosis and treatment were the crucial elements needed to solve them. All the discussed cases were further referred to the law enforcement institutions and were used as expert statements during court trials.

3.1 Case 1: Medical Error

A 36-year old male, who had suffered multiple organ traumas because of traffic accident, was treated within the first several days of hospitalization in the Intensive Care Unit because of multi-organ insufficiency. He needed to be kept sedated because of catecholamine and respiratory therapy prolonged treatment. Head computed tomography allowed identifying serious damages to frontal lobes of the brain cortex and hemorrhagic foci located in deep structures of both brain hemispheres, right side brain chamber hemorrhage and left side subdural hematoma (70 mm in diameter and 6 mm thick). Due to intensive treatment, improvement of the overall status has been achieved – including satisfactory cardiovascular and respiratory capacity. After 20 days, the patient had been moved to Rehabilitation Ward, but was shortly readmitted to the Intensive Care Unit because of acute hyperventilation-related respiratory insufficiency. Bronchoscope examination revealed 2 cm long trachea narrowing that was related to prior prolonged intubation. The necessary surgical treatment was successful, but on the 30th day after the accident, the patient needed a T-type tracheal prosthesis to be implanted, to support, and secure mechanically the desired tracheal lumen diameter. Until the hospitalization in the Thoracic Surgery Ward, there were no reported alterations in the patient's behavior – especially any that would point to the central nervous system located damage. In the ward the patient was described as "aggressive because of central nervous system damage… with very limited reaction to sedatives… because of increasing aggression directed against both himself and the surroundings, the patient, after prior phone consultation, was sent to the Psychiatric Ward to undergo further treatment". Mucolytic drug administration and a regular suction of a T-type airway prosthesis was indicated. The witnesses, while stating in front of the court, pointed to both psychic and physical agitation of the patient, which was linked by him to subjective "shortness of breath feeling" that he reported of as "increasing over time." The patient's aggressive actions were focused mainly on the external exit of T-type airway prosthesis and included repeated attempts to remove it. In the Psychiatric Ward the therapy was generally limited to administering several central nervous system targeting drugs, including haloperidol, carbamazepine, and valproic acid. Medical files confirmed that agitation and aggressive behaviors increased over time along with the reported dyspnea. According to the files, the patient was initiating conflicts with the medical personnel and was presenting an aggressive attitude. After 10 days of psychiatric hospitalization, dyspnea increased, which necessitated an otolaryngological consultation. The patient was diagnosed with purulent inflammation and marked mucous substance deposition in the airway prosthesis, with accompanying mucosa swelling and laryngitis. Neuroleptic treatment was continued to suppress behavioral and aggressive behavior that was linked to morphologic central nervous system lesions. Because of limited success of therapy, the dose of haloperidol was significantly raised and hydroxyzine with benzodiazepines were added to the therapy regimen. At the same time, the initial physician's order to regularly apply suction to the airway prosthesis was cancelled and replaced with the request that the patient takes care of the prosthesis by him, which was deemed possible. Even with the intensified pharmacologic treatment, his mental status deteriorated further, with a rapid increase in aggressive behavior episodes. Dyspnea increased and there appeared forced respiration with rhonchi and rales over the lungs. Since doctors were convinced that the worsening of patient's condition was due solely to the brain lesions, the intensity of pharmacotherapy was increased. After 18 days of psychiatric hospitalization, a psychiatrist on duty was called to intervene, because the patient complained of severe shortness of breath and presented with violent agitation. An i.v. bolus of a benzodiazepine was ordered, which resulted in deepening dyspnea, although deeply sedated patient eventually fell asleep. Before midnight, the patient woke up, showing signs of agitation and disorientation, and started to run around the ward in a chaotic way, arguing noisily with other patients – after a short time he reported he could not breathe anymore and developed complete respiratory arrest within minutes, which was quickly followed by asystole-related cardiac arrest. Resuscitation continued for more than 60 min remained ineffective. Autopsy revealed that the lumen of the airway prosthesis was completely blocked with thick mucus deposits – partially hard dried and partially soft. A microscopic examination confirmed the presence of old and recent inflammatory white blood cells infiltrations of, and bacteria masses on, the bronchial epithelial cells. In forensic medicine experts' opinion that was requested by law enforcement institution, behavioral changes were linked with increasing chronic hypoxia caused by progressive obliteration of airway prosthesis. Hypoxia, especially when it is unaccompanied by rapid carbon dioxide retention, can go unnoticed for prolonged times, exacerbating existing central nervous disorders, and some point rapidly triggering their manifestation, or evoking quite new conditions and symptoms – like anxiety, agitation, aggressive behavior, euphoria, or hallucinations.

3.2 Second Case: Civil Law Procedure

In the year 2008, forensic medicine physicians participated in court proceedings concerning a civil case in the context of inheritance. A 78-year old male patient diagnosed with a terminal stage of lung cancer was treated at home following the family doctor's recommendations. In the last weeks of life, the range of therapeutic activities was limited to strictly palliative treatment aimed at ensuring the maximum patient's comfort. This treatment resulted from a failure of intensive repeat oncologic treatment, in the presence of multiple metastatic lesions confirmed in diagnostic imaging of both lungs. Shortly after the patient's death, it turned out that several hours prior the death he called for a family member and a public notary to change his will. Because other members of the family did not witness this event and changes in the inheritance were unexpected for them, they questioned the ability of the patient to express his will properly in the last moments of life.

The public notary explained in front of the court that the patient "was looking exhausted" and that he heard the patient "repeating the statements made by the family member present", which he took as independent statements of the patient and included into the written form of the will, signed in an unreadable form by the patient. The family doctor and the palliative medicine specialist who had been taking care of the patient in the last several months of his life reported that during that time the patient was barely conscious most of the time. He was suffering from severe dyspnea because of massive bilateral pleural effusions, so in the last weeks he was under strong influence of narcotics and sedatives that were administered in increasing doses to lessen his suffering. Because of that, both physicians considered the patient to be fully incapable of making any valid legal statements, including changing his last will. The court ruling invalidated the will altered in the last moments of life. The illness- and drugs-related hypoxia, especially in terminal patients, is often being used to raise questions about patients' ability to properly express their will, especially last will, in the way demanded by law – it is also extremely difficult for the medical experts' teams to decide upon whether or not it shall be trusted.

3.3 Third Case: Penal Law Procedure

In the year 2004, forensic medicine experts were asked by a judge to issue an opinion on the case of 32-years old woman subjected to general anesthesia because of a routine appendectomy surgery. She reported in the post-operation care room that she had been sexually abused by a seemingly medical staff member – wearing a white physician's hospital suit – just after she partially woke up from anesthesia. The patient's complaint was initially distrusted, but a routine lookup at the hospital staff was immediately initiated. The anesthesiologist who took care of the patient during the operation and then accompanied her for some time alone, as required, in the post-operation room was missing – he was reported by witnesses to wear trousers that appeared to be blood-stained in the groin region when he was leaving the post-operation care room. The anesthesiologist was followed by medical staff to his house where the trousers were retrieved. In the end it turned out that he had left the hospital and took hiding in at home where he was trying in a hurry to wash out his trousers in a washing machine. It also was established just before the surgery that the patient was menstruating, so even a crude microscopic examination could confirmed that the blood on the stained trousers was likely hers. Later on, at the biochemistry department the comparative DNA tests of the patient's blood matched the blood specimens derived from the anesthesiologist's trousers; thus, plainly indicating a sexual assault. The physician argued in court that hypoxia and anesthetics used around the surgery time influenced the patient's judgment and accusations against him; the explanations that were rejected by court.

4 Discussion

Hypoxia is often discussed in the medical literature and its pathophysiology is quite well known. It can result, among others, from insufficient oxygen diffusion in the lungs, limited oxygen blood transportation due to anemia, or low oxygen concentration in the air. No matter what causes hypoxia, it always interferes with cellular metabolism and triggers multiple organ dysfunctions. Hypoxic tissues switch into anaerobic metabolism which leads to acidosis. Each organ's sensibility to hypoxia is different; the most susceptible are those with the most active metabolism, like the heart and liver, but the first to be irreversibly damaged by hypoxia is unquestionably the brain. The signs of malfunction and morphology deterioration differ from organ to organ, but they sum up into an interconnected dysfunctional interactions leading to organism's death if they are not early detected and compensated for on time (Barash et al. 2006). A connection between hypoxia and central nervous system malfunction is of particular interest for forensic medicine specialists. Cases of violent deaths have been long studied and described in medical literature, along with their various mechanisms and interconnected links (Stark 2011; Olshaker et al. 2007). Among central nervous system asphyxia-related death cases there are accidents, homicides, and suicides. Acute brain hypoxia often results from blocking the oxygen-rich blood from reaching the brain by applying some type of compression to the neck organs – homicide deaths are often a result of rope or manual strangulation and hanging is the most popular method of committing suicide among men in Poland (Trnka et al. 2013). Brain hypoxia is of critical importance in the forensic investigation of death as it precipitates rapid loss of consciousness and in people attempting to commit suicide it is soon impossible to revert their hanging even when they are not fully suspended by the rope.

Another field of interest in forensic medicine concerning brain hypoxia is autoerotic hypoxia and accidental deaths that result from it. It is reported that increased sexual satisfaction might be achieved in cases when sexual activity coexists with brain hypoxia. It is often abused by males who set up special, sometimes sophisticated mechanisms allowing them to apply pressure to the neck and automatically release it at the moment they start to lose consciousness. The effects of hypoxia similar to those of feeling like being 'on high' achieved by narcotic drugs have been reported in jet pilots and high mountain climbers, but hypoxia accompanying the autoerotic activities is of purposeful and not accidental nature and is referred to as a sort of sexual deviation (Aggrawal 2009).

Old typical scenarios of asphyxia-related deaths may be relatively easy solved by forensic medicine, in contrast to the atypical aspects of hypoxia currently emerging, especially when they are iatrogenic in nature. In the three hypoxia-related cases above outlined, hypoxia coexisted with the actual offence or misdemeanor being evaluated in court. It seems that nowadays law enforcement institutions ought to get more and more interested in hypoxia side effects and their sequelae. These institutions ask for medical experts' opinions which become complex and sometimes difficult to be issued at all, as it is extremely difficult to find out *post mortem* and not in a closely monitored patient, whether or not hypoxia actually existed, how deep it was, and if and what kind of practical effect it really had for the patient's overall medical condition, consciousness, and behavior.

A 36-year old man described above, treated for airway obstruction that caused chronic hypoxia, is a case in point concerning the difficulties in forensic medicine opinions. The patient's behavior was considered normal during his repeated stay at the Intensive Care Unit and his consciousness was not altered. Initially, he had full contact with the surrounding people; including properly consenting to undergo medical procedures in the way demanded by medical law regulations. Later on, however, the patient's psychiatric status deteriorated and he was transferred to a psychiatric ward, where he presented increasingly intensified behavioral changes, such as agitation, anxiety, and unexpected aggressive outbursts against medical staff and other patients. Physicians associated the symptoms with trauma-related lesions located in brain frontal lobes due to a traffic accident. However, chronic hypoxia also results in brain morphological changes and therefore it should be part of differential diagnosis (Corcoran and O'Connor 2013; Yang et al. 2013; LaManna 2007; Dwinell et al. 2000). Forensic investigation confirmed that the patient's behavioral changes were associated with progressing brain hypoxia; mucous and purulent deposits present in the inflamed airways narrowed the airway prosthesis' lumen, making ventilation ineffective. Similar symptoms, but precipitated by a completely different initial mechanism, have been described in medical literature in high mountain climbers who suffer from central nervous hypoxia because of breathing air with inadequate oxygen partial pressure (Arzy et al. 2005; Virués-Ortega et al. 2004; Brugger et al. 1999). In such cases, lack of carbon dioxide retention, which usually accompanies hypoxia in most clinical pathologies, makes it difficult to diagnose hypoxia on the basis of physical examination. The forensic experts analyzing the case had the comfort of being able to confront the autopsy findings against the witnesses' statements and detailed medical files, all of which allowed drawing a comprehensive diagnosis. It was concluded that the psychiatrists failed in efforts to find the cause that was behind the patient's symptoms. Instead they acted schematically, merely intensifying pharmacological sedation, which in fact worsened the clinical status by depressing ventilator drive and consequently exacerbating respiratory insufficiency and hypoxia (Heuss et al. 2012). The patient received an additional dose of midazolam just a short time before his collapse. A tragic death of the 36-year old man was precipitated by the lack of supervision of bronchial tree and airway prosthesis patency and iatrogenic suppression of ventilatory drive and cough reflex, causing a buildup of debris in the airways. The law enforcement institutions needed an answer to the question of whether or not it was possible for the treating physicians to differentiate between the symptoms of brain hypoxia and those attributable to a trauma of frontal brain lobes; the answer to the question was positive in this particular case.

The questions about brain hypoxia and its influence on cognitive functions of the 78-year old man, described in Case 2 was raised in court. It was argued that brain hypoxia could significantly limit the patient's ability to think reasonably, to make informed decisions, and to declare last will. The expert witnesses needed to review the court file, including medical records, medical personnel, and other witnesses' statements. As the patient was treated mainly at home, the availability of blood oxygenation tests was limited. The available data gave a good enough basis to confirm that the man, prior to his death, had not been able to fully understand or control his actions, which is required to make legally valid decisions concerning property. Metastases in the lungs and pleural effusions caused chronic blood acidosis; the available arterial blood gas content revealed: PO_2 of 65 mmHg, PCO_2 of 55 mmHg, and pH of 7.19. The lab tests were consistent with the witnesses' statements and the whole case supports the argument that chronic central nervous hypoxia impairs consciousness and cognitive functions (Wiklund et al. 2012; Wilkinson et al. 2010; Powell et al. 2000). The court shared the forensic opinion and annulled an altered version of the patient's last will.

In contrast to the two cases described above, hypoxia in Case 3 was not a harmful factor that should be considered and was overlooked, but it was taken advantage of by a perpetrator of sexual crime. The question raised by court was about the witness' credibility. Thus, forensic experts were asked to say whether or not the woman's statement about her being sexually abused in the perioperative period could be trusted, which would eliminate the possibility of hypoxia-related hallucinations, secondary to anesthesia. The defense strategy of the accused of sexual crime anesthesiologist was based on medical knowledge, since it is rather impossible to fully eliminate such an option based only on clinical data. On the other hand, the patient's story was consistent with the physical evidence, including the woman's menstruation blood present on the anesthesiologist's trousers and medical staff's statements. Consequently, the patient's accusation was trusted and the anesthesiologist was considered guilty of sexual assault, even in the face of medical reports describing hallucinations triggered by anesthetics, like ketamine or propofol, anesthesia-related brain hypoxia (Friedberg 1993; Perel and Davidson 1976; Fine and Finestone 1975).

5 Conclusions

Reports of hypoxia influencing human mental status raise interest in law enforcement institutions and courts. The lawyers try to use hypoxic states to both accuse and defend their clients. There is also more demand for medical forensic experts in such cases, but experience, scientific, and empirical basis to judge and resolve such issues is still limited. Central nervous system hypoxia and functional changes secondary to it are going to be considered as factors of potential importance also in cases that do not fit into asphyxia scenarios of forensic medicine.

Conflicts of Interest The authors declare no conflicts of interest in relation to this article.

References

Aggrawal A (2009) Forensic and medico-legal aspects of sexual crimes and unusual sexual practices. CRC Press, Boca Raton, pp 201–247

Arzy S, Idel M, Landis T, Blanke O (2005) Why revelations have occurred on mountains? Linking mystical experiences and cognitive neuroscience. Med Hypotheses 65:841–845

Barash PG, Cullen BF, Stoelting RK (2006) Clinical anesthesia, 5th edn, Chapter 39. Lippincott Williams & Wilkins, Philadelphia, pp 2259–2260

Brugger P, Regard M, Landis T, Oelz O (1999) Hallucinatory experiences in extreme-altitude climbers. Neuropsychiatry Neuropsychol Behav Neurol 12:67–71

Corcoran A, O'Connor JJ (2013) Hypoxia-inducible factor signalling mechanisms in the central nervous system. Acta Physiol 208:298–310

Dwinell MR, Huey KA, Powell FL (2000) Chronic hypoxia induces changes in the central nervous system processing of arterial chemoreceptor input. Adv Exp Med Biol 475:477–484

Fine J, Finestone SC (1975) Sensory disturbances following ketamine anesthesia: recurrent hallucinations. Anesth Analg 52:428–430

Friedberg BL (1993) Hypnotic doses of propofol block ketamine-induced hallucinations. Plast Reconstr Surg 91:196–197

Heuss LT, Sugandha SP, Beglinger C (2012) Carbon dioxide accumulation during analgosedated colonoscopy: comparison of propofol and midazolam. World J Gastroenterol 38:5389–5396

LaManna JC (2007) Hypoxia in the central nervous system. Essays Biochem 43:139–151

Olshaker JS, Jackson MC, Smock WS (2007) Forensic emergency medicine, 2nd edn. Lippincott Williams & Wilkins, Philadelphia, pp 85–146, 249–260

Perel A, Davidson JT (1976) Recurrent hallucinations following ketamine. Anaesthesia 31:1081–1083

Powell FL, Huey KA, Dwinell MR (2000) Central nervous system mechanisms of ventilatory

acclimatization to hypoxia. Respir Physiol 121:223–236
Stark MM (2011) Clinical forensic medicine – a physician's guide, 3rd edn. Springer, New York, pp 71–132
Trnka J, Gęsicki M, Suslo R, Siuta J, Drobnik J, Pirogowicz I (2013) Death as a result of violent asphyxia in autopsy reports. Adv Exp Med Biol 788:413–416
Virués-Ortega J, Buela-Casal G, Garrido E, Alcázar B (2004) Neuropsychological functioning associated with high-altitude exposure. Neuropsychol Rev 14:197–224
Wiklund L, Martijn C, Miclescu A, Semenas E, Rubertsson S, Sharma HS (2012) Central nervous tissue damage after hypoxia and reperfusion in conjunction with cardiac arrest and cardiopulmonary resuscitation: mechanisms of action and possibilities for mitigation. Int Rev Neurobiol 102:173–187
Wilkinson KA, Huey K, Dinger B, He L, Fidone S, Powell FL (2010) Chronic hypoxia increases the gain of the hypoxic ventilatory response by a mechanism in the central nervous system. J Appl Physiol 109:424–430
Yang Q, Wang Y, Feng J, Cao J, Chen B (2013) Intermittent hypoxia from obstructive sleep apnea may cause neuronal impairment and dysfunction in central nervous system: the potential roles played by microglia. J Neuropsychiatr Dis Treat 9:1077–1086

Advs Exp. Medicine, Biology - Neuroscience and Respiration (2015) 6: 57–66
DOI 10.1007/5584_2014_47

Published online: 14 October 2014

Does Health Status Influence Acceptance of Illness in Patients with Chronic Respiratory Diseases?

D. Kurpas, B. Mroczek, J. Brodowski, M. Urban, and A. Nitsch-Osuch

Abstract

The level of illness acceptance correlates positively with compliance to the doctor's recommendations, and negatively with the frequency and intensity of complications of chronic diseases. The purpose of this study was to determine the influence of the clinical condition on the level of illness acceptance, and to find variables which would have the most profound effect on the level of illness acceptance in patients with chronic respiratory diseases. The study group consisted of 594 adult patients (mean age: 60 ± 15 years) with mixed chronic respiratory diseases, recruited from patients of 136 general practitioners. The average score in the Acceptance of Illness Scale was 26.2 ± 7.6. The low level of illness acceptance was noted in 174 (62.6 %) and high in 46 (16.6 %) patients. Analysis of multiple regressions was used to examine the influence of explanatory variables on the level of illness acceptance. The variables which shaped the level of illness acceptance in our patients included: improvement of health, intensity of symptoms, age, marital status, education level, place of residence, BMI, and the number of chronic diseases. All above mentioned variables should be considered during a design of prevention programs for patients with mixed chronic respiratory diseases.

D. Kurpas (✉)
Department of Family Medicine, Medical University of Wroclaw, 1 Syrokomli St., 51-141 Wroclaw, Poland

Institute of Nursing, Public Higher Medical Professional School, Opole, Poland
e-mail: dkurpas@hotmail.com

B. Mroczek
Department of Humanities in Medicine, Faculty of Health Sciences, Pomeranian Medical University, 48 Zolnierska St., 70-204 Szczecin, Poland

J. Brodowski
Laboratory of General Practice, Faculty of Health Sciences, Pomeranian Medical University, Szczecin, Poland

M. Urban
Great Poland Center of Pulmonology and Thoracic Surgery "Eugenia and Janusz Zeyland", 62 Szamarzewskiego St., 60-569 Poznan, Chodziez Hospital (branch), 64-800 Chodziez

A. Nitsch-Osuch
Department of Family Medicine, Medical University of Warsaw, Warsaw, Poland

Keywords
Adaptation to disease • Chronic care model • Chronic disease • Prophylaxis • Somatic status

1 Introduction

The future of primary care is closely related to challenges issued by patients with at least two chronic diseases, i.e., the patients who require systematic contacts with a general practitioner (GP), additional medical examinations, and regular administration of drugs. Most important is an increase in the level of illness acceptance and all of these factors result in the continuously improving quality of life (QoL). Thus, the efficient care provided by a multidisciplinary team should be based on the knowledge of the particular population structure and the encouragement of patients to co-decide about further management of their diseases (with emphasis on exchange of experiences) as well as the process of education oriented toward the patients' own resources, and the evaluation program of: adverse actions, compliance to doctor's recommendations, unsuccessful therapy, and the recurrence of the complaints (Gerlach et al. 2006; Wagner 2000). Such a model of health care includes bio-psycho-social factors, which should be analyzed from the point of view of individual needs and attitudes demonstrated by chronically ill patients (Flores et al. 2007). The quality assessment of medical care offered within the above model (Chronic Care Model) consists of the evaluation of the process (work of interdisciplinary team taking care of chronically ill patients and their families) and the evaluation of the structure (staff and equipment in a GP's surgery). These aspects should be assessed with reference to the result of medical care, which is defined by the level of illness acceptance (Kringos et al. 2010; Donabedian 1980).

A high level of illness acceptance is regarded to be a factor that supports health, since it shows that a person is able to adapt to the disease-related limitations and accepts disability. It also reflects the level of dependence on other people (Juczynski 2009). Illness acceptance is extremely important in case of patients with chronic respiratory diseases, who have to stay within the health care structures continuously (Wright and Kirby 1999). Adaptation to a chronic disease decreases negative reactions and emotions associated with it and its therapy. The higher the level of illness acceptance by a patient, the lower the number of destructive reactions and emotions should be expected (Fortin et al. 2006). Adaptation to a chronic illness, manifest by a high level of illness acceptance, is essential for the effective disease control, which includes the patient's involvement in self-control. It is a critical element of health care provided for patients with chronic respiratory diseases (Nowicki and Ostrowska 2008).

Lickiewicz et al. (2010) argue that the level of illness acceptance correlates positively with compliance to doctor's recommendations concerning the diagnostic and therapeutic processes, and negatively with the frequency and intensity of complications of chronic diseases. Patients with higher acceptance of chronic diseases react positively to the instructions for therapy, easier accept the role of a patient, and deal with the limitations caused by a chronic disease more constructively while using their own resources (Lickiewicz et al. 2010). The study of Juczynski and Adamiak (2000) showed that sometimes there is no direct relationship between QoL and the intensity of clinical symptoms of a disease. Illness acceptance modulates QoL – even in cases of clinically advanced chronic diseases – and a high level of illness acceptance can increase the level of QoL.

The purpose of this study was to determine the influence of the clinical condition on the level of illness acceptance, and to find variables which have the most profound effects on the level of illness acceptance in patients with chronic respiratory diseases.

2 Methods

The study was carried out in accordance with The Code of Ethics of the World Medical Association (Declaration of Helsinki) and was approved by the Bioethical Commission of the Medical University of Wroclaw (no. KB 608/2011). The main inclusion criteria were: age (at least 18 years), the diagnosis of at least one chronic disease, and written informed consent to take part in the study.

The study group consisted of 594 adult patients with chronic respiratory diseases. The mean age was 60 ± 15 years. Detailed sociodemographic data of the patients are shown in Table 1, and the list of chronic diseases diagnosed in Table 2. Participants of the study were recruited from the patients of 136 general practitioners between July 2011 and April 2013. The physicians presented patients with a diagnosis of a chronic respiratory disease. The project staff members asked the eligible patients if they would like to take part in the study. Patients who agreed to participate anonymously in the project signed a consent form. The patients were given a questionnaire to complete at home and to return in a stamped envelope.

The patients' adaptation to life with a disease was assessed using the Acceptance of Illness Scale developed by Felton et al. (1984), and adapted to the Polish conditions by Juczynski (2009). The scale consists of eight statements about negative consequences of health state, where every statement is rated on a five-point Likert scale, where one denotes poor adaptation to a disease and 5 its full acceptance. The score for illness acceptance is a sum of all points and can range from 8 to 40. Low scores (0–29) indicate the lack of acceptance and poor adaptation to a disease, and a strong feeling of mental discomfort. High scores (35–40) indicate the acceptance of illness, manifest as the lack of negative emotions associated with a disease. The scale can be used to assess the degree of acceptance of every disease. The α-Cronbach coefficient of the Polish version is 0.85 (Juczynski 2009) and that of the original version is 0.82 (Felton et al. 1984).

Table 1 Socio-demographic data of the chronically ill patients

	n	%
Gender		
Women	148	53.2
Men	130	46.8
Age		
≤24	11	6.1
25–44	25	9.0
45–64	111	39.9
65–84	112	40.3
≥85	13	4.7
Place of residence		
Village	97	34.9
Town/city population		
< 5,000	37	13.3
5,000–10,000	10	3.6
10,000–50,000	64	23.0
50,000–100,000	39	14.0
100,000–200,000	17	6.1
Over 200,000	14	5.0
Education		
Primary	62	22.3
Vocational	84	30.2
Secondary	64	23.0
Post-secondary	37	13.3
Higher	31	11.2
Marital status		
Single	25	9.0
Married	182	65.5
Divorced	14	5.0
Widowed	57	20.5

The clinical condition of patients was determined by means of a questionnaire measuring the improvement of somatic and psychic health, somatic symptoms occurring most frequently, and the results of spirometry and laboratory tests during the past 12 months. For the sake of analysis, the somatic index was calculated for each patient. Somatic symptoms reported by the patients were assigned a value from one (symptoms occurring once a year) to seven (permanent symptoms). The index was calculated by

Table 2 Patients' diagnoses according to the 2013 ICD-10-CM Diagnosis Codes and comorbidity ($n = 278$)

	n	%
Diagnosis[a]		
J45 Bronchial asthma	130	46.8
J44 Other chronic obstructive pulmonary diseases	86	30.9
J42 Unspecified chronic bronchitis	39	14.0
J41 Chronic simple and mucous-purulent bronchitis	35	12.6
J43 Pulmonary emphysema	29	10.4
J47 Bronchiectasis	24	8.6
Most common co-existing diseases[a]		
I10 Primary hypertension	75	27.0
M47 Spondylosis	58	20.9
I70 Atherosclerosis	39	14.0
E11 Type 2 diabetes mellitus	26	9.4
I50 Heart failure	26	9.4
I11 Hypertensive heart disease	24	8.6
M15 Osteoarthritis of multiple joints	24	8.6
I25 Chronic ischemic heart disease	22	7.9
E10 Type 1 diabetes mellitus	15	5.4
K58 Irritable bowel syndrome	15	5.4
Comorbidity		
1 chronic disease	56	20.1
2 and more	222	79.9
3 and more	172	61.9
4 and more	112	40.3
5 and more	65	23.4
6 and more	39	14.0
7 and more	25	9.0
8 and more	14	5.0
9 and more	9	3.2
10 and more	5	1.8
11 and more	3	1.1
15 and more	2	0.7

[a]Some patients were diagnosed as having at least two pathological entities

summing up the values assigned to somatic symptoms, and then dividing the sum by 49 (the highest possible score for the frequency of somatic symptoms).

2.1 Statistical Elaboration

Statistical analysis was performed using R 2.10.1 (for Mac OS X Cocoa GUI). The type of distribution for all variables was determined. Overall, the results of variables were not normally distributed, which was confirmed by the Shapiro-Wilk normality test. Arithmetic means, standard deviations, medians, as well as the range of variability (extremes) were calculated for measurable (quantitative) variables, while for qualitative variables, the frequency (percentage) was determined. For each variable pair, the Spearman rank correlation coefficient (r) was calculated. The null hypothesis was tested that the correlation coefficient is 0, which would denote no correlation between variables.

The analysis of multiple regressions was used in order to examine the impact of explanatory variables on the level of illness acceptance, somatic index, and on the subjective improvement of the somatic and psychic health of patients. An initial set of explanatory variables for each response variable included only these variables that significantly correlated with a particular response variable. All available models with different numbers of explanatory variables were checked. Ultimately, 13 significant models which met the assumptions of multiple regressions were found; namely, 8 models with 4 explanatory variables for the response variable 'level of illness acceptance' and 5 models with 3 explanatory variables for the response variable 'somatic index'. No model was found for the response variable 'subjective improvement of the somatic and psychic health'. The critical level of significance was assumed at $p < 0.05$.

3 Results

During the past 12 months preceding the study, the improvement of somatic health was reported by 193 (69.4 %) patients, while no improvement was reported by 85 (30.6 %) patients. The improvement of psychic health was reported by 194 (69.8 %) patients, and no improvement was found by 84 (30.2 %) patients. The average value of the somatic index in the study group was 0.4 ± 0.2, average forced expiratory volume in 1 s (FEV1) was 75.0 ± 12.7 l, serum creatinine 0.9 ± 0.5 mg/dl, total cholesterol 219.2 ± 43.8 mg/dl, and glucose 125.4 ± 54.3 mg/dl. The most

Table 3 Top 10 medications used by patients

Active substance	*n*	%
Formoterol fumarate	57	20.5
Theophylline	45	16.2
Acetylsalicylic acid	43	15.5
Salbutamol	39	14.0
Ipratropium bromide	31	11.2
Salmeterol	27	9.7
Fluticason propionate	25	9.0
Ramipril	21	7.6
Budesonide	19	6.8
Furosemide	18	6.5

common complaints were: permanent pain of the spine (39, 29.8 %) and peripheral joints (47, 33.3 %), high blood pressure (above 140/90 mmHg) 2–3 times a week (56, 29.0 %), permanent dyspnea (72, 28.4 %), chest pain 2–3 times a week (52, 27.2 %), stomachache once a month (27, 21.6 %). Most patients (175, 62.9 %) had abnormal body mass: overweight (85, 30.6 %), obesity (81, 29.1 %), and underweight (9, 3.2 %). The Body Mass Index (BMI) of 191 (41.9 %) patients was within normal range.

The average number of chronic diseases was 3.4 ± 2.2 and that of medications taken by patients was 4.4 ± 3.0. The top 10 medications used by the patients are presented in Table 3. The average score in the Acceptance of Illness Scale was 26.2 ± 7.6. The low level of illness acceptance (score of 1–29 points) was noted in 174 (62.6 %), medium (30–34 points) in 58 (20.9 %), and high (35–40 points) in 46 (16.6 %) patients.

3.1 Results of Correlations

More men than women lived in rural areas and the majority of women lived in places with a population of over 200,000 citizens ($r = -0.12$, $p = 0.048$). Incomplete primary education was more common among widows/widowers and graduate education more common among unmarried women/men ($r = -0.20$, $p = 0.001$).

Patients in advanced age were more often widows/widowers ($r = 0.51$, $p < 0.0001$), had incomplete primary education ($r = -0.48$, $p < 0.0001$), and lived in rural areas ($r = 0.37$, $p < 0.0001$). This group of patients more often had high BMI values ($r = 0.13$, $p = 0.032$), a higher number of chronic diseases ($r = 0.52$, $p < 0.0001$), no improvement of somatic ($r = -0.23$, $p < 0.0001$) and psychic ($r = -0.28$, $p < 0.0001$) health during the past 12 months, high values of the somatic index ($r = 0.41$, $p < 0.0001$), used a higher number of medications ($r = 0.22$, $p < 0.001$), and had abnormal creatinine levels ($r = -0.54$, $p = 0.008$).

Patients with incomplete primary education (*vs.* graduate education) more often lived in rural areas ($r = -0.56$, $p < 0.0001$), and had a higher number of chronic diseases ($r = -0.36$, $p < 0.0001$), no improvement of somatic ($r = 0.13$, $p = 0.037$) and psychic ($r = 0.16$, $p = 0.009$) health during the past 12 months, higher values of the somatic index ($r = -0.30$, $p < 0.0001$), and high FEV1 values ($r = -0.49$, $p = 0.043$).

Widows/widowers (*vs.* unmarried women/men) more often had high BMI values ($r = 0.19$, $p = 0.001$), a higher number of chronic diseases ($r = 0.36$, $p < 0.0001$), no improvement of somatic ($r = -0.17$, $p = 0.006$) and psychic ($r = -0.16$, $p = 0.008$) health during the past 12 months, high values of the somatic index ($r = 0.19$, $p = 0.001$), and severe asthma or chronic obstructive pulmonary disease (COPD) as assessed by spirometry ($r = -0.60$, $p = 0.012$).

Patients from rural areas (*vs.* patients from cities/towns with a population over 200,000) more often had a higher number of chronic diseases ($r = 0.18$, $p = 0.003$), no improvement of somatic ($r = -0.14$, $p = 0.017$) and psychic ($r = -0.13$, $p = 0.032$) health during the past 12 months, and high values of the somatic index ($r = 0.16$, $p = 0.010$).

The number of chronic diseases was higher among widows/widowers ($r = 0.36$, $p < 0.0001$), patients with incomplete primary education ($r = -0.36$, $p < 0.0001$), patients from rural areas ($r = 0.18$, $p = 0.003$), advanced age patients ($r = 0.52$, $p < 0.0001$), patients with no improvement of somatic

($r = -0.29$, $p < 0.0001$) and psychic ($r = -0.25$, $p < 0.0001$) health during the past 12 months, patients with high values of the somatic index ($r = 0.60$, $p < 0.0001$), patients with high BMI values ($r = 0.26$, $p < 0.0001$), patients taking a lot of medications ($r = 0.27$, $p < 0.0001$), and patients with abnormal total cholesterol levels ($r = 0.39$, $p = 0.047$).

Patients who felt that their somatic health improved during the past 12 months more often had an improvement of psychic health ($r = 0.63$, $p < 0.0001$), low values of the somatic index ($r = -0.29$, $p < 0.0001$), high levels of illness acceptance ($r = 0.23$, $p < 0.0001$), and took a low number of medications ($r = -0.17$, $p = 0.001$). No improvement of somatic health was more common among advanced age patients ($r = -0.23$, $p < 0.0001$), widows/widowers ($r = -0.17$, $p = 0.006$), patients with incomplete primary education ($r = 0.13$, $p = 0.037$), patients from rural areas ($r = -0.14$, $p = 0.017$), patients with a higher number of chronic diseases ($r = -0.29$, $p < 0.0001$), and those using a lot of medications ($r = -0.16$, $p = 0.008$).

Lack of improvement of psychic health during the past 12 months was more common among patients in advanced age ($r = -0.28$, $p < 0.0001$), widows/widowers ($r = -0.16$, $p = 0.008$), patients with incomplete primary education ($r = 0.16$, $p = 0.009$), patients from rural areas ($r = -0.13$, $p = 0.032$), patients with a higher number of chronic diseases ($r = -0.25$, $p < 0.0001$), patients with high values of the somatic index ($r = -0.27$, $p < 0.0001$), and those using a higher number of medications ($r = -0.17$, $p = 0.005$).

High values of the somatic index were more often observed among patients in advanced age ($r = 0.41$, $p < 0.0001$), widows/widowers ($r = 0.19$, $p = 0.001$), patients with incomplete primary education ($r = -0.30$, $p < 0.0001$), and patients from rural areas ($r = 0.16$, $p = 0.010$). Patients with high values of the somatic index more often had high BMI values ($r = 0.14$, $p = 0.016$), used a higher number of medications ($r = 0.15$, $p = 0.010$), had normal serum cholesterol levels ($r = 0.52$, $p = 0.006$), and had abnormal creatinine levels ($r = -0.39$, $p = 0.010$).

Low FEV1 values were more common among patients with graduate education ($r = -0.49$, $p = 0.043$), and those below 24 years of age ($r = 0.54$, $p = 0.026$). The third degree asthma and stage 3 COPD were more often observed in unmarried women/men (*vs.* widows/widowers) ($r = -0.60$, $p = 0.012$). Normal cholesterol levels were more often observed among patients with a higher number of chronic diseases ($r = 0.39$, $p = 0.047$), and high values of the somatic index ($r = 0.52$, $p = 0.006$).

Patients with high BMI values more often took a higher number of medications ($r = 0.10$, $p = 0.025$). High BMI values were more common in patients with a high number of chronic diseases ($r = 0.26$, $p < 0.0001$).

Patients using a higher number of medications more often had abnormal creatinine levels ($r = -0.35$, $p = 0.022$). A higher number of medications was more common among patients in advanced age ($r = 0.22$, $p < 0.001$), with a higher number of chronic diseases ($r = 0.27$, $p < 0.0001$), without improvement of somatic ($r = 0.16$, $p = 0.008$) and psychic ($r = -0.17$, $p = 0.005$) health during the past 12 months, and with high values of the somatic index ($r = 0.15$, $p = 0.010$).

Low levels of illness acceptance were more often observed among patients with high BMI values ($r = -0.15$, $p = 0.014$), a high number of chronic diseases ($r = -0.47$, $p < 0.0001$), without improvement of somatic ($r = 0.30$, $p < 0.0001$) and psychic ($r = 0.26$, $p < 0.0001$) health during the past 12 months, with high values of the somatic index ($r = -0.42$, $p < 0.0001$), with low total cholesterol levels ($r = 0.41$, $p = 0.038$), those with high serum glucose levels ($r = -0.47$, $p = 0.017$), widows/widowers ($r = -0.22$, $p < 0.0001$), patients with incomplete primary education ($r = 0.39$, $p < 0.0001$), patients from rural areas ($r = -0.28$, $p < 0.0001$), and patients in advanced age ($r = -0.51$, $p < 0.0001$).

3.2 Multiple Regression Results

The results of multiple regression are presented in Table 4. An increase in illness acceptance was associated with eight variable groupings listed below, in the order from the most to least influential explanatory variable:

1. improvement of somatic health during the past 12 months, living in a city with a population over 200,000 citizens, being single, a low BMI;
2. younger age, higher education, improvement of somatic health during the past 12 months, a low number of chronic diseases;
3. younger age, higher education, a low number of chronic diseases, improvement of psychic health during the past 12 months;
4. younger age, higher education, a low value of the somatic index, improvement of somatic health during the past 12 months;
5. higher education, a low value of the somatic index, improvement of somatic health during the past 12 months, a low number of chronic diseases;
6. higher education, a low value of the somatic index, a low number of chronic diseases, improvement of psychic health during the past 12 months;
7. low value of the somatic index, a low number of chronic diseases, living in a city with a population over 200,000, improvement of somatic health during the past 12 months;
8. low value of the somatic index, a low number of chronic diseases, living in a city with a population over 200,000, improvement of psychic health during the past 12 months.

A decrease in the somatic index was associated with five variable groupings listed below, in the order from the most to least influential explanatory variable:

1. improvement of psychic health during the past 12 months, higher education, being single;
2. improvement of somatic health during the past 12 months, a low BMI, being single;
3. improvement of psychic health during the past 12 months, being single, a low BMI;

Table 4 Results of multiple regression

Explanatory variables	Estimate coefficient	SE	t-value	Pr (>|t|)	Multiple r^2	Adjusted r^2	p for F-test
Response variable: level of illness acceptance							
Model 1							
Improvement of somatic health	0.257	0.056	4.563	<0.0001	0.168	0.156	<0.0001
Place of residence	−0.201	0.056	−3.607	<0.001			
Marital status	−0.119	0.057	−2.093	0.037			
BMI	−0.111	0.056	−1.986	0.048			
Model 2							
Age	−0.312	0.060	−5.225	<0.0001	0.360	0.350	<0.0001
Education	0.226	0.055	4.147	<0.0001			
Improvement of somatic health	0.170	0.051	3.350	<0.001			
Number of chronic diseases	−0.140	0.055	−2.553	0.011			
Model 3							
Age	−0.320	0.061	−5.286	<0.0001	0.343	0.333	<0.0001
Education	0.220	0.055	3.978	<0.0001			
Number of chronic diseases	−0.158	0.055	−2.861	0.005			
Improvement of psychic health	0.103	0.052	1.993	0.047			
Model 4							
Age	−0.307	0.058	−5.256	<0.0001	0.369	0.359	<0.0001
Education	0.216	0.054	3.983	<0.0001			
Somatic index	−0.174	0.054	−3.233	0.001			
Improvement of somatic health	0.161	0.051	3.189	0.002			

(continued)

Table 4 (continued)

Explanatory variables	Estimate coefficient	SE	t-value	Pr (>|t|)	Multiple r^2	Adjusted r^2	p for F-test
Model 5							
Education	0.320	0.052	6.192	<0.0001	0.320	0.310	<0.0001
Somatic index	−0.190	0.061	−3.102	0.002			
Improvement of somatic health	0.179	0.053	3.418	<0.001			
Number of chronic diseases	−0.148	0.061	−2.450	0.015			
Model 6							
Education	0.314	0.053	5.985	<0.0001	0.304	0.293	<0.0001
Somatic index	−0.204	0.062	−3.299	0.001			
Number of chronic diseases	−0.161	0.061	−2.630	0.009			
Improvement of psychic health	0.119	0.053	2.249	0.025			
Model 7							
Somatic index	−0.228	0.063	−3.594	<0.001	0.258	0.247	<0.0001
Number of chronic diseases	−0.188	0.063	−2.979	0.003			
Place of residence	−0.186	0.053	−3.530	<0.001			
Improvement of somatic health	0.164	0.055	2.974	0.003			
Model 8							
Somatic index	−0.236	0.064	−3.709	<0.001	0.248	0.237	<0.0001
Number of chronic diseases	−0.196	0.063	−3.097	0.002			
Place of residence	−0.191	0.053	−3.594	<0.001			
Improvement of psychic health	0.123	0.055	2.229	0.027			
Response variable: somatic index							
Model 1							
Improvement of psychic health	−0.213	0.057	−3.691	<0.001	0.127	0.118	<0.0001
Education	−0.198	0.058	−3.446	<0.001			
Marital status	0.118	0.058	2.037	0.043			
Model 2							
Improvement of somatic health	−0.260	0.058	−4.522	<0.0001	0.115	0.106	<0.0001
BMI	0.132	0.057	2.297	0.022			
Marital status	0.123	0.058	2.111	0.036			
Model 3							
Improvement of psychic health	−0.239	0.058	−4.137	<0.0001	0.105	0.095	<0.0001
Marital status	0.128	0.059	2.180	0.030			
BMI	0.127	0.058	2.199	0.029			
Model 4							
Improvement of somatic health	−0.260	0.056	−4.642	<0.0001	0.145	0.136	<0.0001
Education	−0.212	0.056	−3.762	<0.001			
BMI	0.130	0.056	2.321	0.021			
Model 5							
Improvement of psychic health	−0.230	0.057	−4.050	<0.0001	0.130	0.120	<0.0001
Education	−0.204	0.057	−3.561	<0.001			
BMI	0.127	0.057	2.240	0.026			

4. improvement of somatic health during the past 12 months, higher education, a low BMI;
5. improvement of psychic health during the past 12 months, higher education, a low BMI.

4 Discussion

In the present study the most common sustained symptoms in patients with chronic respiratory diseases were pain complaints, which have been also found by other studies in heterogeneous groups of chronically ill patients (Agborsangaya et al. 2013). A lower number of somatic symptoms (lower somatic index) were observed among patients in advanced age, widows/widowers, patients with higher education, and those living in rural areas, which may result from the fact that these patients go to a doctor quicker than others, as soon as the first symptoms appear (Cianciara 2012). The results demonstrate that an increase in somatic symptoms should be expected if these patients also show no improvement of their psychic and somatic health during the past 12 months, are obese, have primary education, and are single.

Most common problems with the Chronic Care Model implementation (Helseth et al. 1999; O'Connor 1998) include the situations when a patient has a low level of knowledge concerning the complications of chronic diseases (Anderson et al. 1991, 1992) or is not convinced that the doctor's recommendations will produce the desired results (Helseth et al. 1999), and professionals are convinced that a patient has no will to change his lifestyle (Golin et al. 1996; Wing et al. 1985; Christensen et al. 1983). Clinical observations show that education programs based on patients' own resources are more effective in overcoming these barriers than simply giving information (Bodenheimer et al. 2002). However, it is not possible to achieve this goal without the sufficient level of chronic disease acceptance, which results from appropriate education (Nowicki and Ostrowska 2008; Ouwens et al. 2005; Wagner 2000).

The majority of the analyzed patients with chronic respiratory diseases demonstrated low level of illness acceptance. Nevertheless, the average score of illness acceptance of 26.2 ± 7.6 we obtained was higher than those observed by Juczynski (2009) in dialysed patients (25.3 ± 6.0), men after a myocardial infarction (22.1 ± 6.1), patients with multiple sclerosis (24.6 ± 7.2), patients with chronic neuropathic pain (18.5 ± 7.1), patients with spinal pain (20.5 ± 8.7), women with migraine (24.2 ± 7.7), and patients with type 2 diabetes (24.8 ± 7.1). In that study the patient groups were less numerous and were homogeneous regarding the type of chronic disease. On the other hand, we found that the level of illness acceptance in our patients with chronic respiratory diseases was lower than that noted by Juczynski (2009) among women with cancer of breast and uterus (28.1 ± 7.6); suggesting that this group of patients participated in more effective education and received psychological support, which results in better illness acceptance (Nowicki and Ostrowska 2008).

An asymptomatic character of chronic diseases, diagnostics based on the analysis of existing symptoms, and the lack of immediate treatment effects show that patients often do not undertake the therapy or interrupt it, do not turn up for check-ups, and do not undergo diagnostic tests. All these factors result in postponed therapy and early complications. The improvement in this sphere can be achieved through an increase in illness acceptance, and thus higher QoL in chronically ill patients (Nowicki and Ostrowska 2008; Wasowski and Marcinowska-Suchowierska 2006; Wright and Kirby 1999).

According to the presented results, the prevention program oriented toward increasing the acceptance of illness should be addressed to patients with chronic respiratory diseases who do not show any improvement of their somatic and psychic health, whose somatic symptoms intensified during the past 12 months, and who have more than three co-morbidities, are obese, are of advanced age, live in rural areas, are widows/widowers, and have primary education.

5 Conclusions

Subjective assessment of the somatic and psychic health improvement and the intensity of somatic symptoms shape the level of illness acceptance among patients with chronic respiratory diseases. Other variables contributing to level of illness acceptance in patients with chronic respiratory diseases include: age, marital status, the level of education, the place of residence, the BMI, and the number of chronic diseases. All mentioned variables should be considered during a design process of prevention programs for patients with mixed chronic respiratory diseases.

Conflicts of Interest The authors have no financial or otherwise relations that might lead to a conflict of interest.

References

Agborsangaya CB, Lau D, Lahtinen M, Cooke T, Johnson JA (2013) Health-related quality of life and healthcare utilization in multimorbidity: results of a cross-sectional survey. Qual Life Res 22:791–799

Anderson RM, Donnelly MB, Dedrick RF, Gressard CP (1991) The attitudes of nurses, dietitians, and physicians toward diabetes. Diabetes Educator 17:261–268

Anderson RM, Donnelly MB, Davis WK (1992) Controversial beliefs about diabetes and its care. Diabetes Care 15:859–863

Bodenheimer T, Wagner EH, Grumbach K (2002) Improving primary care for patients with chronic illness. JAMA 288:1775–1779

Christensen NK, Terry RD, Wyatt S, Pichert JW, Lorenz RA (1983) Quantitative assessment of dietary adherence in patients with insulin-dependent diabetes mellitus. Diabetes Care 6:245–250

Cianciara D (2012). Siemens report. Available from: www.siemens.pl/pool/healthcare/raport_siemensa_2012.pdf. Accessed 15 Jan 2013

Donabedian A (1980) Explorations in quality assessment and monitoring, vol 1, The definition of quality and approaches to its assessment. Health Administration Press, Ann Arbor

Felton BJ, Revenson TA, Hinrichsen GA (1984) Stress and coping in the explanation of psychological adjustment among chronically ill adults. Soc Sci Med 18:889–898

Flores LM, Davis R, Culross P (2007) Community health: a critical approach to addressing chronic diseases. Prev Chronic Dis 4:A108

Fortin M, Bravo G, Hudon C, Lapointe L, Dubois MF, Almirall J (2006) Psychological distress and multimorbidity in primary care. Ann Fam Med 4:417–422

Gerlach FM, Beyer M, Muth C, Saal K, Gensichen J (2006) New perspectives in the primary care of the chronically ill–against the 'tyranny of the urgent'. German J Evid Qual Health Care 100:335–343

Golin CE, DiMatteo MR, Gelberg L (1996) The role of patient participation in the doctor visit: implications for adherence to diabetes care. Diabetes Care 19:1153–1164

Helseth LD, Susman JL, Crabtree BF, O'Connor PJ (1999) Primary care physicians' perceptions of diabetes management: a balancing act. J Fam Pract 48:37–42

Juczynski Z (2009) Measurement instruments in promotion and psychology of health. Laboratory of Psychological Tests of the Polish Psychological Society, Warsaw (in Polish)

Juczynski Z, Adamiak G (2000) Psychological and behavioral predictors of the quality of life of people with multiple sclerosis. Pol Merkur Lekarski 8:413–415

Kringos DS, Boerma WG, Hutchinson A, van der Zee J, Groenewegen PP (2010) The breadth of primary care: a systematic literature review of its core dimensions. BMC Health Serv Res 10:65

Lickiewicz B, Zwolinska-Wcislo M, Lickiewicz J, Rozpondek P, Mach T (2010) Significance of personality features in the illness adaptation process in patients with inflammatory bowel disease. Przeglad Gastroenterologiczny 3:157–163

Nowicki A, Ostrowska Z (2008) Disease acceptance in patients after surgery from breast cancer during supplementary treatment. Pol Merkur Lekarski 24:403–407

O'Connor PJ (1998) From blame to understanding: moving diabetes care forward. J Fam Pract 46:205–206

Ouwens M, Wollersheim H, Hermens R, Hulscher M, Grol R (2005) Integrated care programmes for chronically ill patients: a review of systematic reviews. International J Qual Health Care 17:141–146

Wagner EH (2000) The role of patient care teams in chronic disease management. Br Med J 320:569–572

Wasowski M, Marcinowska-Suchowierska E (2006) Reasons for failure of pharmacological treatment of chronic diseases with a special emphasis on osteoporosis. Postepy Nauk Medycznych 6:359–366

Wing RR, Epstein LH, Nowalk MP, Scott N, Koeske R (1985) Compliance to self monitoring of blood glucose: a marked-item technique compared with self-report. Diabetes Care 8:456–460

Wright SJ, Kirby A (1999) Deconstructing conceptualizations of 'adjustment' to chronic illness: a proposed integrative framework. J Health Psychol 4:259–272

Advs Exp. Medicine, Biology - Neuroscience and Respiration (2015) 6: 67
DOI 10.1007/5584_2014

Index

Zeitfracht Medien GmbH
Ferdinand-Jühlke-Straße 7
99095 Erfurt, Deutschland
produktsicherheit@kolibri360.de